International Conference on Sustainable, Renewable & Energy Efficiency 2022

AA007313

IOP Conference Series: Materials Science and Engineering
Volume 1278

Bangi, Malaysia
7 – 8 November 2022

ISBN: 978-1-7138-6984-9
ISSN: 1757-8981

Printed from e-media with permission by:

Curran Associates, Inc.
57 Morehouse Lane
Red Hook, NY 12571

Some format issues inherent in the e-media version may also appear in this print version.

Additional copies of this publication are available from:

Curran Associates, Inc.
57 Morehouse Lane
Red Hook, NY 12571 USA
Phone: 845-758-0400
Fax: 845-758-2633
Email: curran@proceedings.com
Web: www.proceedings.com

TABLE OF CONTENTS

PREFACE

The inaugural ***International Conference on Sustainable, Renewable & Energy Efficiency*** (ICSREE 2022) is a conference organized by the Solar Energy Research Institute (SERI), Universiti Kebangsaan Malaysia (UKM). The primary aim of this conference is to create a holistic and hybrid platform for researchers from various backgrounds to share and learn. Besides that, a collaboration with the University of Nottingham, UK, on a special workshop of ***Innovation in Sustainable Cooling Technology***, supported by the British Council, Malaysian Industry-Government Group for High Technology (MIGHT), and the Royal Society, is also being held as a part of this conference. Our sincerest hope is that this conference will inspire current and future generations to embrace and embark on the green and clean energy era.

On behalf of the publication committees, we would like to extend our acknowledgment and gratitude for the efforts and support of the committee members, including academic staffs, administrative staffs, researchers, and postgraduate students for all their efforts and contributions. As the leading institute in renewable and green technology in Malaysia, we firmly believe that this conference will serve as a dynamic springboard for all the students and researchers involved toward critical and innovative thinking. We also hope that this conference inspires them in improving their research skills and will lead towards higher impact research in the future.

Thank you very much.

Ts. Dr. Puvaneswaran Chelvanathan
Chairman / Editors
ChM. Dr. Ubaidah Syafiq / Dr. Masita Mohammad
Publication Committees / Editors
International Conference on Sustainable, Renewable & Energy Efficiency (ICSREE 2022)
Solar Energy Research Institute (SERI)
Universiti Kebangsaan Malaysia
43600 Bangi, Selangor
MALAYSIA
Website: www.ukm.my/seri
Email: eventseri@gmail.com

IOP Conf. Series: Materials Science and Engineering 1278 (2023) 011001 doi:10.1088/1757-899X/1278/1/011001

ORGANIZING COMMITTEES

	Committee	Name
1	Advisor	Associate Prof. Dr. Mohd Adib Ibrahim
2	Chairperson	Ts. Dr. Puvaneswaran Chelvanathan
3	Co-Chairpersons	Associate Prof. Dr. Adnan Ibrahim Dr. Hasila Jarimi
4	Secretaries	Mrs. Siti Aminah Bahron Ms. Alya Athirah Ahmad
5	Treasury	ChM. Sharifah Nurain Syed Nasir Mrs. Nur Idatul Diana Ahmad Zaki
6	Publicity	Dr. Halim Razali Dr. Md Khan Sobayel Rafiq Mr. Muhamad Samsuri Samsudin
7	Program and Schedule	Dr. Ahmad Fazlizan Abdullah Dr. Kazi Sajedur Rahman Mr. Amirul Iman Zainudin
8	Technical	Ir. Ts. Dr. Mohd Faizal Fauzan Mr. Asmadi Hazim Mr. Mohamad Arif Mokhtar Mr. Raja Mohd Saleh Raja Hasan Mr. Mohamad Ridzuan Md Razi
9	Publication	ChM. Dr. Ubaidah Syafiq Dr. Masita Mohammad Mr. Ahmad Afif Safwan Mohd Radzi
10	Scientific	Prof. Dato' Ts. Dr. Kamaruzzaman Sopian Prof. Dr. Md Akhtaruzzaman Associate Prof. Dr. Mohd Asri Mat Teridi Associate Prof. Dr. Norasikin Ahmad Ludin Associate Prof. Dr. Nilofar Asim Ar. Dr. Lim Chin Haw ChM. Dr. Mohd Sukor Su'ait Dr. Ahmad Fudholi

IOP Conf. Series: Materials Science and Engineering 1278 (2023) 011002 doi:10.1088/1757-899X/1278/1/011002

Peer Review Statement

All papers published in this volume have been reviewed through processes administered by the Editors. Reviews were conducted by expert referees to the professional and scientific standards expected of a proceedings journal published by IOP Publishing Publishing.

- **Type of peer review:** Single Anonymous
- **Conference submission management system:** Morressier
- **Number of submissions received:** 30
- **Number of submissions sent for review:** 28
- **Number of submissions accepted:** 17
- **Acceptance Rate (Submissions Accepted / Submissions Received × 100):** 56.7
- **Average number of reviews per paper:** 1.18
- **Total number of reviewers involved:** 17
- **Contact person for queries:**
  Name: Ubaidah Syafiq
  Email: ubaidahsyafiq@ukm.edu.my
  Affiliation: Solar Energy Research Institute - The National University of Malaysia (SERI-UKM)

IOP Conf. Series: Materials Science and Engineering 1278 (2023) 012001 doi:10.1088/1757-899X/1278/1/012001

Comprehensive Modelling for Analyzing the Power Conversion Efficiency of Polycrystalline Silicon Photovoltaic Device under Indoor Operating Conditions

Son Qian Liew, Kok-Keong Chong*

Lee Kong Chian Faculty of Engineering and Science, Universiti Tunku Abdul Rahman, Bandar Sungai Long, 43000 Kajang, Selangor, MALAYSIA

Abstract. Recently, most of the studies were mainly focused on the variation between outdoor solar spectral irradiance and standard AM1.5G spectrum, as well as their effects towards the performances of different PV technologies. Interestingly, limited studies have been conducted to study the effect of artificial lighting on the power conversion efficiency of the photovoltaic (PV) device. The spectral irradiance produced by indoor lighting differs significantly as compared to the that of standard AM1.5G spectrum, which inevitably leads to different performances of the PV under standard test conditions and under artificial lighting exposure. In this study, the performance modelling has been carried out based on the comprehensive method proposed by Chong et al. to predict the PCE of organic solar cell devices. From the measurement results, the spectrum of LED illumination has been compared with that of AM1.5G, in which the LED shows a relatively narrow band with significant peaks at 450 and 560 nm. Furthermore, the theoretical model has been successfully validated by comparing the simulated PCE and the measured PCE at reasonably deviations in the range of 3.0 – 23.3%.

* *Corresponding author:* chongkk@utar.edu.my, kokkeong_c@yahoo.com

1. Introduction

The body-centric internet of things (IoT) market is expected to encompass tens of billions of linked devices by 2020, thanks to the proliferation of self-contained low-power indoor electrical devices with wireless internet connections [1–3]. Conventional batteries are cumbersome and expensive in many body-centric sensors for IoT applications because they are much larger than tiny devices. Therefore, it is desirable to have technologies that can power these electronics using only indoor energy, without additional batteries or electrical wiring. Indoor photovoltaic (PV) modules with high energy densities and relatively large output voltages have been shown to be the most appropriate energy sources in previous investigations of available power-generating devices in indoor lighting situations [4]. Therefore, in this work, the measurement of *J–V* curve of PV has been proposed to analyze its performance under the illumination of the light emitting diode (LED), in which the irradiance level is usually around 0.2 – 1.0% of the outdoor solar irradiance.

Based on the literature, there were minimal studies on the methodology to predict PV performance under weak indoor irradiances (1 – 50 Wm^{-2}). Instead, most of the studies focused on predicting PV efficiency under standard test conditions (STC) at AM1.5G, 1000 Wm^{-2}, operating temperature of 25 ℃, which can differ significantly from the LED illumination. The outputs of PV devices may show a remarkable variation under various indoor operating conditions. Therefore, a theoretical model that can predict PV performance under LED illumination is vital for the design of indoor PV applications.

In this study, we would like to validate the theoretical model proposed by Chong et al. for predicting the performance of PV under different LED intensities. A few parameters were taken as inputs for the model to achieve the prediction, i.e., external quantum efficiency (EQE), *J–V* curves, and LED spectral irradiances [5–6]. For the validation of the model, the simulated power conversion efficiency (PCE) is compared with actual measurements under indoor conditions.

2. Methodology

The analytical methodology to predict the PCE under varying LED spectral irradiances has been implemented to evaluate the performance of polycrystalline silicon PV. Then, the modelled results are compared with the measured PCE under LED illumination. This theoretical model necessitates a series of computational processes, including various inputs: i.e., *J–V* characteristics for different intensities, the measured LED spectrum, and the EQE of the tested PV.

The initial step of the process is to acquire *J-V* curves of the PV devices under various LED intensities. A few parameters can be extracted from the measured *J–V* curves, such as short-circuit current density (J_{sc}), open-circuit voltage (V_{oc}), and fill factor (FF). The extracted results are summarized and plotted as V_{oc} vs. J_{sc} and FF vs. J_{sc} graphs to obtain the relationship of V_{oc} and FF in the function of J_{sc}. Shockley equation is used to determine the relationship between V and J:

$$J = \frac{R_p}{R_s+R_p}\left\{J_s\left[\exp\left(\frac{q(V-JR_s)}{nkT}\right)-1\right]+\frac{V}{R_p}\right\}-J_{ph}(V) \tag{1}$$

where the R_p is the device parallel resistance, R_s is the device series resistance, q is the electron charge, n is the device ideality factor, k is the Boltzmann constant, T is the temperature in Kelvin and J_{ph} is the voltage-dependent photocurrent density.

According to the Shockley equation, V_{oc} can be derived as the following given that $J = 0$, $V = V_{oc}$ and $J_{sc} = J_{ph}(0) >> J_s$:

$$V_{oc}=\frac{nkT}{q}\ln(J_{sc})+\left[V_{oc1}-\frac{nkT}{q}\ln(J_{sc1})\right] \tag{2}$$

where the series of values for V_{oc1} and J_{sc1} can be extracted from the *J–V* plot.

On the other hand, the values of FF and PCE can be obtained using the formulas as shown below:

$$\text{FF} = \frac{J_{max} \times V_{max}}{J_{sc} \times V_{oc}} \tag{3}$$

$$\text{PCE} = \frac{J_{sc} \times V_{oc} \times \text{FF} \times A_{cell}}{P_{in}} \tag{4}$$

where J_{max} and V_{max} represent maximum current density ($mAcm^{-2}$) and maximum voltage (V) at the maximum power output point of the *J–V* curve respectively. A_{cell} represents the active area (m^2) of the polycrystalline silicon PV module, which is exposed to the LED illumination during the *J–V* curve measurement. P_{in} represents the input power of the incident LED light with the unit of Watt.

The irradiance of the LED illumination, I_{in}, can be calculated using the measured spectral irradiance obtained from the AVANTES spectrometer, where the related relationship is presented by equation (5). On the other hand, the spectral responsivity, $R(\lambda)$ is defined in equation (6).

$$I_{in} = \frac{P_{in}}{A} = \int_{\lambda_1}^{\lambda_2} S_L(\lambda)\, d\lambda \tag{5}$$

$$R(\lambda) = \frac{q\lambda}{hc} \cdot \eta_{EQE}(\lambda) \tag{6}$$

where I_{in} represents the irradiance (Wm^{-2}), $S_L(\lambda)$ represents the measured spectral irradiance of LED, λ_1 and λ_2 are the lower (300 nm) and the higher (1700 nm) limits of the wavelength of LED illumination, respectively. Last but not least, J_{sc} can be computed by integrating $S_L(\lambda)$ and $R(\lambda)$ throughout all the wavelengths, as shown in the following equation:

$$J_{SC} = \int_{\lambda_1}^{\lambda_2} S_L(\lambda) \cdot R(\lambda)\, d\lambda \tag{7}$$

3. Results and discussion

The peak intensities and energy distribution of spectral irradiances, $S_L(\lambda)$, usually vary significantly when the different types of light sources are considered. Hence, the spectral irradiance of LED was measured using the AVANTES spectrometers to cover the wavelength ranging from 300 to 1700 nm, which is comprised of AvaSpec-2048-USB2-RM (visible bandwidth 200–1100 nm) and AvaSpec-NIR256-1.7-RM (infrared bandwidth 1000–1700 nm). The corresponded results are presented in **Figure 1**, where the blue colour line represents the AM1.5G spectrum, whilst red line represents the LED illumination of 9.3 Wm^{-2}. Unlike the $S_L(\lambda)$ graph of the AM1.5G, the $S_L(\lambda)$ graph of the LED as shown in **Figure 1** is a combination of a sharp peak at approximately 400 – 500 nm superimposed on a broader peak at the wavelength range of approximately 500 – 700 nm.

A set of current-density versus voltage (J-V) curves for the polycrystalline silicon PV device were acquired by varying the intensities of LED ranging from 2.3 to 9.3 Wm^{-2}. The corresponding results are presented in **Figure 2**, while the associated PV parameters are listed in **Table 1**. As expected, both open-circuit voltage (V_{oc}) and short-circuit current density (J_{sc}) decrease with decreasing light intensity. The measured PCE of the PV device under the intensity of 9.3 Wm^{-2} is the highest (PCE = 10.0%), with J_{sc} of 3.58 $mAcm^{-2}$ and V_{oc} of 5.3 V. On the other hand, the output of PV device measured at the lowest intensity (2.3 Wm^{-2}) exhibits the lowest PCE of 6.5%, with J_{sc} of 0.90 $mAcm^{-2}$ and V_{oc} of 4.09 V. The total number of photons absorbed by the polycrystalline silicon cell decreases when the light intensity decreases, directly affecting the short circuit current of the PV device. Continuity to that, the external quantum efficiency (EQE) explains the phenomenon, in which, when the light intensity is decreased, the number of effective carriers decreases and hence the measured J_{sc} decreases [7].

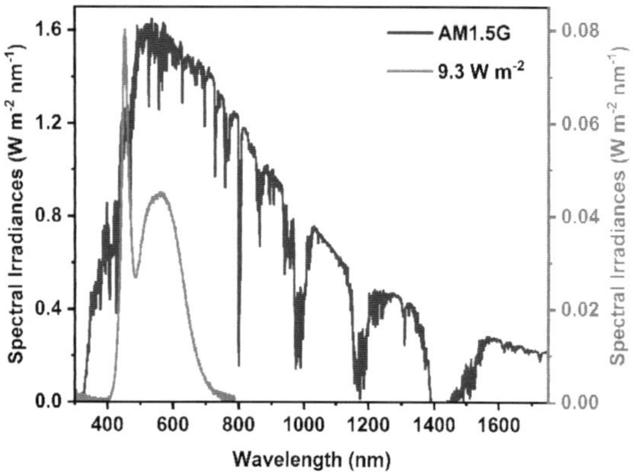

Figure 1. Spectral irradiances of AM1.5G and LED at the intensity of 9.3 Wm^{-2}.

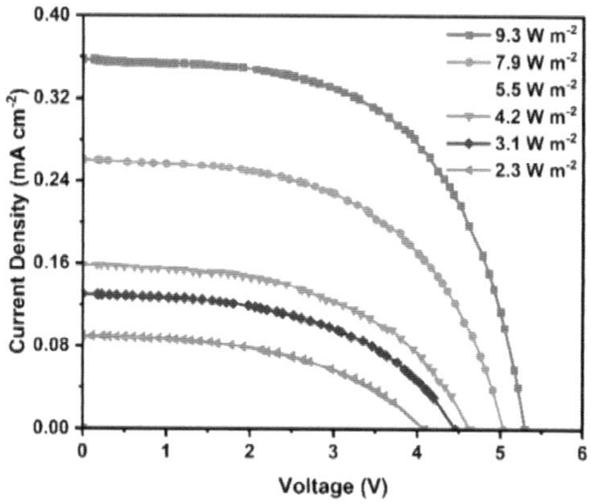

Figure 2. *J–V* characteristics of polycrystalline silicon photovoltaic device under different intensities of LED illumination.

Table 1. The parameters of the polycrystalline silicon photovoltaic device based on the indoor measurement: V_{oc}, J_{sc}, FF, PCE under LED illumination at different intensities.

I_{in} (W/m^2)	Indoor measurement			
	V_{oc} (V)	J_{sc} (mAcm^{-2})	FF	PCE (%)
9.3	5.30	3.58	0.59	10.0
7.9	5.04	2.60	0.55	7.6
5.5	4.88	2.12	0.53	8.3
4.2	4.63	1.59	0.51	7.5
3.1	4.46	1.30	0.50	7.9
2.3	4.09	0.90	0.48	6.5

As aforementioned, EQE is considered in the theoretical modelling process. Therefore, it is vital to investigate the effective range of the EQE that can produce a meaningful output. As such, the spectral irradiance of LED under different intensities are compared with the EQE of the polycrystalline silicon PV, as shown in **Figure 3**. By comparing the effective region of the EQE ranging from 300 to 850 nm with the spectral irradiances of the LED, first half of the region for LED spectral irradiances fall into declining trend of the EQE. Besides that, the input power of the incident LED light, P_{in}, can be computed by integrating the area below the spectral irradiance graphs plotted in Figure 3, which is significant in the range of 400 – 750 nm. As predicted, the peak height of the spectrum is decreased when the intensity of LED is decreased.

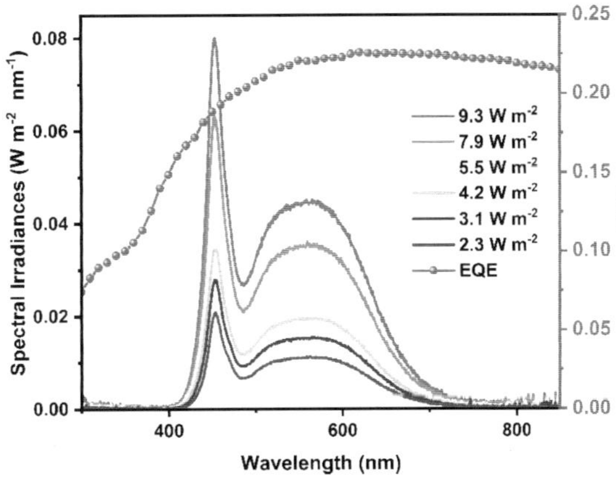

Figure 3. LED spectrums of the LED at different intensities (2.3 – 9.3 Wm^{-2}) and the EQE of the polycrystalline silicon PV.

The next part of the work was to apply the comprehensive theoretical modelling to compute the theoretical PCE of the polycrystalline silicon PV under different intensities of LED. The measured parameters were used to obtain the correlations for V_{oc} vs. J_{sc} and FF vs. J_{sc} under different LED intensities as shown in **Figure 4** and **Figure** 5, respectively. From the graphs, it can be deduced that both regression lines closely fit the data, where R^2 values are obtained as 0.998 and 0.948 for the V_{oc} vs. J_{sc} graph and FF vs. J_{sc} graph, respectively. Therefore, the resulting equations can generate accurate results for the theoretical model. The empirical formulas for V_{oc} vs. J_{sc} and FF vs. J_{sc} were generated using the curve fitting function as presented in the following:

$$V_{oc} = 0.8692\ln(J_{sc}) + 6.2127 \tag{8}$$

$$\mathrm{FF} = 0.0755\ln(J_{sc}) + 0.6549 \tag{9}$$

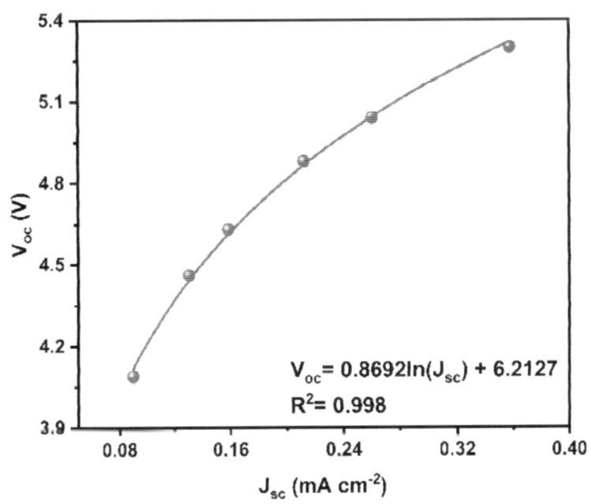

Figure 4. The plots for V_{oc} vs. J_{sc}, under different LED irradiances, as extracted from the *J–V* characteristic plots.

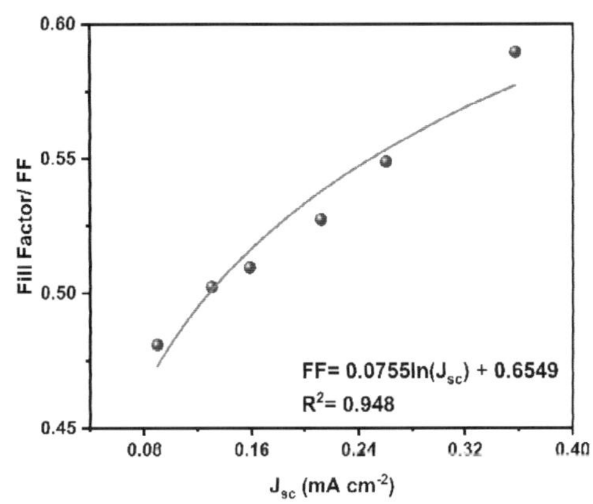

Figure 5. The plots for FF vs. J_{sc}, under different LED irradiances, as extracted from the *J–V* characteristic plots.

The theoretical modelled PV parameters, i.e., V_{oc}, J_{sc}, FF, and PCE, were calculated and summarized in **Table 2**. It can be deduced that the theoretical modelled PV parameters have similar trend as those of actual measured parameters. The computed values for V_{oc}, J_{sc}, and FF were based on equations (2), (7), and (3), accordingly. The measured PCE reveals a decreasing trend as the intensity of LED is reduced from 9.3 to 2.3 Wm^{-2}. From the Figure 3, the spectral distribution of LED is not reduced in the proportional manner when the intensity of LED illumination is decreased, which has caused more serious mismatch between EQE of the PV module and the spectral irradiances of LED as the intensity of LED is lower. The relatively higher portion energy distribution of LED shifts to lower range wavelength (400-500 nm) can reduce the output energy because the EQE drops sharply for the wavelength from 500 nm to 300 nm, which has subsequently affected the PCE of PV module. Previous works by Chong et al. and Hong et al. mainly utilized numerical analysis to predict the PCE of the PV device under various solar irradiances other than that of AM1.5G. However, in this work, the proposed theoretical model was to predict the performance of the PV under LED light illumination. To investigate the deviation between the measured and theoretical modelled values, the ΔPCE and ΔJ_{sc} were computed using equation (10) and equation (11), respectively.

$$\Delta\text{PCE}(\%) = \frac{\text{measured PCE} - \text{simulated PCE}}{\text{measured PCE}} \times 100\% \quad (10)$$

$$\Delta J_{sc}(\%) = \frac{\text{measured } J_{sc} - \text{simulated } J_{sc}}{\text{measured } J_{sc}} \times 100\% \quad (11)$$

Table 2. Photovoltaic parameters of the polycrystalline silicon photovoltaic device based on the theoretical modelling: V_{oc}, J_{sc}, FF, PCE under LED illumination at different intensities.

I_{in} (W/m^2)	Theoretical modelled parameters			
	V_{oc} (V)	J_{sc} ($mAcm^{-2}$)	FF	PCE (%)
9.3	5.30	3.58	0.59	10.0
7.9	5.17	3.02	0.56	9.3
5.5	4.82	2.00	0.53	7.9
4.2	4.59	1.54	0.51	7.3
3.1	4.32	1.13	0.49	6.4
2.3	4.05	0.83	0.47	5.8

The calculated values for ΔPCE and ΔJ_{sc} were summarized and listed in **Table 3**. Based on the data, the values for ΔPCE and ΔJ_{sc} are in the range of 2.8 – 22.2% and 2.8 – 26.3% respectively. The variation between theoretical and experimental results are considered as random error, which mainly caused by the measurement error, i.e., the uncertainty of the lux meter that is used to measure the LED intensities. In short, the prediction is reasonably accurate for LED lighting conditions (low light intensity).

Table 3. ΔPCE and ΔJ_{sc} between the actual and modelled parameters.

I_{in} (W/m^2)	ΔPCE	ΔJ_{sc}
9.3	–	–
7.9	22.2	15.8
5.5	5.8	26.3
4.2	2.8	2.8
3.1	18.4	13.6
2.3	11.3	7.8

4. Conclusion

In this work, the *J–V* curves of the polycrystalline PV device have been measured under different LED irradiances. The measured PCEs under LED lighting show a decreasing trend when the intensities of LED illumination are decreased. The spectrum of LED illumination has been investigated and compared with the spectrum of AM1.5G, in which the LED shows a relatively narrow band with significant peaks at 450 and 560 nm. The empirical formulas were computed for V_{oc} vs. J_{sc} and FF vs. J_{sc}, which served as input values for the theoretical modelling. Finally, the theoretical modelling has been successfully validated by comparing the between the measured and the simulated PCE values for different LED intensities in the range of 2.3 – 9.3 Wm^{-2}, whereby the prediction is reasonably accurate with ΔPCE in the range of 2.8 – 22.2%.

5. Acknowledgements

This research was supported by the Ministry of Higher Education (MoHE) Malaysia through Fundamental Research Grant Scheme project FRGS/1/2021/TK0/UTAR/01/1. Besides, the authors would also like to express their gratitude to UTAR Strategic Research Fund 2018 with project number IPSR/RMC/UTARSRF/PROJECT 2018-C1/001 (vote account 6274/011) for the financial support.

References

[1] Lee J G *et al.*, 2015 D2ART: Direct Data Accessing from Passive RFID Tag for infra-less, contact-less, and battery-less pervasive computing *Microprocess. Microsyst.* **39**, 8 p. 767–781.

[2] Gubbi J Buyya R Marusic S and Palaniswami M, 2013 Internet of Things (IoT): A vision, architectural elements, and future directions *Futur. Gener. Comput. Syst.* **29**, 7 p. 1645–1660.

[3] Miorandi D Sicari S De Pellegrini F and Chlamtac I, 2012 Internet of things: Vision, applications and research challenges *Ad hoc networks* **10**, 7 p. 1497 1516.

[4] Yin H *et al.*, 2020 Thick-Film Low Driving-Force Indoor Light Harvesters *Sol. RRL* **4**, 10 p. 2000291.

[5] Chong K K Khlyabich P P Hong K J Reyes-Martinez M Rand B P and Loo Y L, 2016 Comprehensive method for analyzing the power conversion efficiency of organic solar cells under different spectral irradiances considering both photonic and electrical characteristics *Appl. Energy* **180** p. 516–523.

[6] Hong K J Tan S T Chong K-K Yap C C Jumali M H H and Loo Y-L, 2019 Numerical analysis with experimental verification to predict outdoor power conversion efficiency of inverted organic solar devices *Renew. Energy* **135** p. 589–596.
[7] Li Z Yang J and Dezfuli P A N, 2021 Study on the Influence of Light Intensity on the Performance of Solar Cell *Int. J. Photoenergy* **2021**.

IOP Conf. Series: Materials Science and Engineering 1278 (2023) 012002 doi:10.1088/1757-899X/1278/1/012002

Numerical Simulation of Ultrathin CdTe Solar Cell by SCAPS-1D

Bibi Zulaika Bhari[1], Kazi Sajedur Rahman[1]*, Puvaneswaran Chelvanathan[1], Mohd Adib Ibrahim[1].

[1]Solar Energy Research Institute (SERI), Universiti Kebangsaan Malaysia, 43600 Bangi, Selangor, Malaysia.

*Corresponding Author: sajed@ukm.edu.my

Abstract. Cadmium telluride (CdTe) has been recognized as one of the efficient and cost-effective thin film solar cell materials with a direct band energy of 1.5 eV. CdTe has long been a leading material in fabrication of solar cell due to its high optical absorption coefficient and ideal band gap. Despite the advantages of CdTe PV technology, CdTe is facing the challenge of Te scarcity. However, it is possible to decrease the CdTe thickness without much compromise in efficiency. Reducing the absorber layer thickness can lower the cost and usage of materials. It can assist to produce large scale CdTe solar cell module as Te is not an earth-abundant element. Numerical simulation of thin film solar cell is a crucial process for defining the possibility of anticipated solar structures, predicting the impact of differences in material characteristics and geometry on overall efficiency. In this research, Solar Cell Capacitance Simulator (SCAPS-1D) is applied to explore the impact of absorber layer thickness and carrier concentration in realizing ultrathin CdTe solar cell. It has been found that 500 nm thick absorber layer is sufficient for acceptable range of cell efficiency. Simulation results of 500 nm CdTe with the carrier concentration of 1.0×10^{14} cm^{-3} has an efficiency of 2.2%. For CdTe carrier concentration of 1.0×10^{17} cm^{-3}, the efficiency increases to 13.22% with open-circuit voltage of 0.988 V, a short-circuit current density of 16.19 mA/cm^2 and fill factor of 82.54%. The optimal numerical solar cell design suggests an approach to further enhance the efficiency of CdTe solar cells.

1. Introduction

CdTe has drawn substantial attention due to its efficiency devices and cost ratio because of its facile flexibility to large scale manufacturing process. Due to its characteristics, CdTe has been regarded as the most suitable material for producing high-efficiency solar cells [1]. Besides, CdTe has direct band gap (~1.45 eV) with absorption coefficient (>10^5 cm^{-1}) which fit to solar cell spectrum and can absorb more than 90% of incoming photon at thickness of 1000 nm [2][3]. A transparent conductive oxide (TCO), CdS film, CdTe absorber, and back contact are the typical layers that make up a superstrate CdTe solar cell's conventional structure.

Generally, the thickness of absorber layer is around 4000 nm. As the absorber layer become thinner, it desires to have low crystallinity thin films with pin hole. To avoid the pin holes by reaching through the window layer, thicker absorber layer is used which can lead to shorting from the back contact. Due to its characteristic, a significant portion of incident photons with energies greater than the bandgap will be absorbed in CdTe layer within a few microns. As a result, less absorber layer is needed than other material. If this thickness can be lessened to a minimum value without considerable weakening in performance, the amount of material used will be cut, shortening a manufacturing time and its cost as

IOP Conf. Series: Materials Science and Engineering 1278 (2023) 012002 doi:10.1088/1757-899X/1278/1/012002

well [4][5]. This study is focused on ultrathin CdTe solar cell, with CdTe thickness less than 1 micron. However, some researcher discovered that it is possible to decrease the absorber layer without much deal in efficiency. In addition, tellurium (Te) resources can be reduced to the potential human harm of cadmium ions, and it is abundant in all Earth seafloor. Reducing the amount of CdTe absorber layer used in solar cell would, from the perspective of the industry, not only reduced material cost and extend the supply of Te, which has a finite reserve, but also lessen the environmental issue associated with disposing of a potentially hazardous material. Furthermore, with less CdTe thickness, it can help to improve the understanding of material issue and CdS/CdTe device operation. Based on previous research, an ultrathin solar cell has the challenges from a few main issues, which are incomplete photo absorption and degradation in efficiency [6]. CdS layer defined the quantity of photons to be attracted to the absorber layer, the quantity of electron-hole pairs to be produced, while enhancing the short circuit current. As the thickness of CdTe is reduced, it leads to a gradual decrease in performances. Much of the efficiency loss is due to incomplete photo absorption. Additionally, when the thickness of the absorber layer is decreased, the effect will be worse in cell performances because of shunting, a fully depleted absorber layer, or interference between the main and back contact junctions when CdTe deals with a particular limit. The advantage of CdTe are they can absorb the sun energy at shorter wavelength and produce a high energy density.

To study the physical performance of solar cells, a numerical modelling is used to examine its capability. In this work, SCAPS-1D was utilized to examine the performances of the solar cell. Several researchers have used various parameters such as carrier lifetime, carrier density, back/front contact, the thickness of layer and doping concentration to scrutinize the performances of the solar cell [7][8]. In this study, absorber layer thickness and carrier concentration are varied to enhance the performance. The key focus in this simulation is to analyze an ultrathin CdTe absorber layer for solar cell applications.

2. Numerical Modelling

Table 1 represents the material parameters used as input during simulation. The simulation is conducted under the illumination of AM1.5G with intensity of 100 mW/cm^2. The drift-diffusion method is employed in this simulation, which essentially relies on the mathematical solution of Poisson's equation and the continuity equations for free holes and electrons [9]. Figure 1 demonstrates the schematic diagram for conventional CdTe solar cell structure (ITO/CdS/CdTe). While modelling, simulation parameters are proposed, and their values are selected based on theoretical analysis and literature reviews.

Table 1. Material parameters used in SCAPS-1D simulator.

Parameter	**ITO**	**CdS**	**CdTe**
Thickness (nm)	185	50	500 - 2500
Band gap (eV)	3.6	2.4	1.5
Electron affinity (eV)	4.8	4.5	4.0
Dielectric permittivity	9	10	10
Conduction band density of states (cm^{-3})	2.0x10^{17}	2.2x10^{18}	8.0x10^{17}
Valence band density of states (cm^{-3})	1.8x10^{19}	1.8x10^{19}	1.8x10^{19}
Electron mobility (cm^2/Vs)	50	100	320
Hole mobility (cm^2/Vs)	5	50	40
Donor/Acceptor carrier density (cm^{-3})	1.0x10^{16}	1.0x10^{16}	10^{13} - 10^{17}

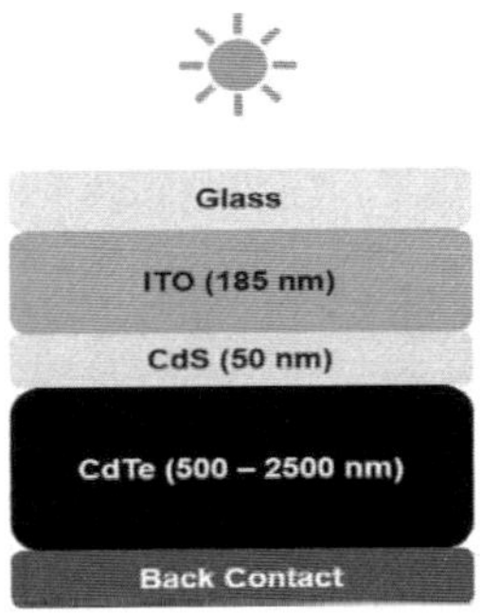

Figure 1. Schematic diagram of conventional CdTe solar cell structure.

3. Results and Discussion

3.1. Impact of CdTe absorber layer thickness

The absorber layer thickness is an influencing parameter of the solar cell structure. CdTe absorber layer thickness is adjusted from 500 nm to 2500 nm to analyze the influences on photovoltaic solar cell parameters such as J_{sc}, FF, V_{oc}, and efficiency. Figure 2 displays the effects of the absorber layer thickness. As the thickness of the absorber layer increases, short circuit current (J_{sc}) and fill factor (FF) increase marginally. V_{oc} relies on a few parameters including light intensity, temperature, charge carrier mobility and bandgap. Meanwhile, V_{oc} slightly rises as the CdTe thickness increases which is due to the effective enhancement of holes mobility. Thus, the reliance of V_{oc} on J_{sc} and J_o, is given by [10]:

$$V_{OC} = \frac{nKT}{q} \ln \left(\frac{Jsc}{Jo} + 1\right) \tag{1}$$

Whereby K and T represent the Boltzmann constant and temperature. As illustrated in Figure 2, J_{sc} increases as the thickness rises because the active layer absorbs more photons. The decrease in current density results due to the less production of electron–hole pair as less number of electron–hole pair can reach the absorber layer. Equation 1 states that a lower J_o is necessary for a higher V_{oc}.

The increase in the thickness of the absorber layer has the propensity to absorb light. Light will be absorbed into the absorber layer more readily as it becomes thicker. Consequently, additional electron-hole pairs will be produced [11]. In addition, as the thickness increases, the number of photoelectrons also increases. Hence, the recombination interface decreases, and the width of the depletion region increases [12]. A CdTe cell with an efficiency higher than 13.22% is found with a 500 nm thick CdTe absorber layer which illustrates the prospect of ultrathin CdTe solar cells.

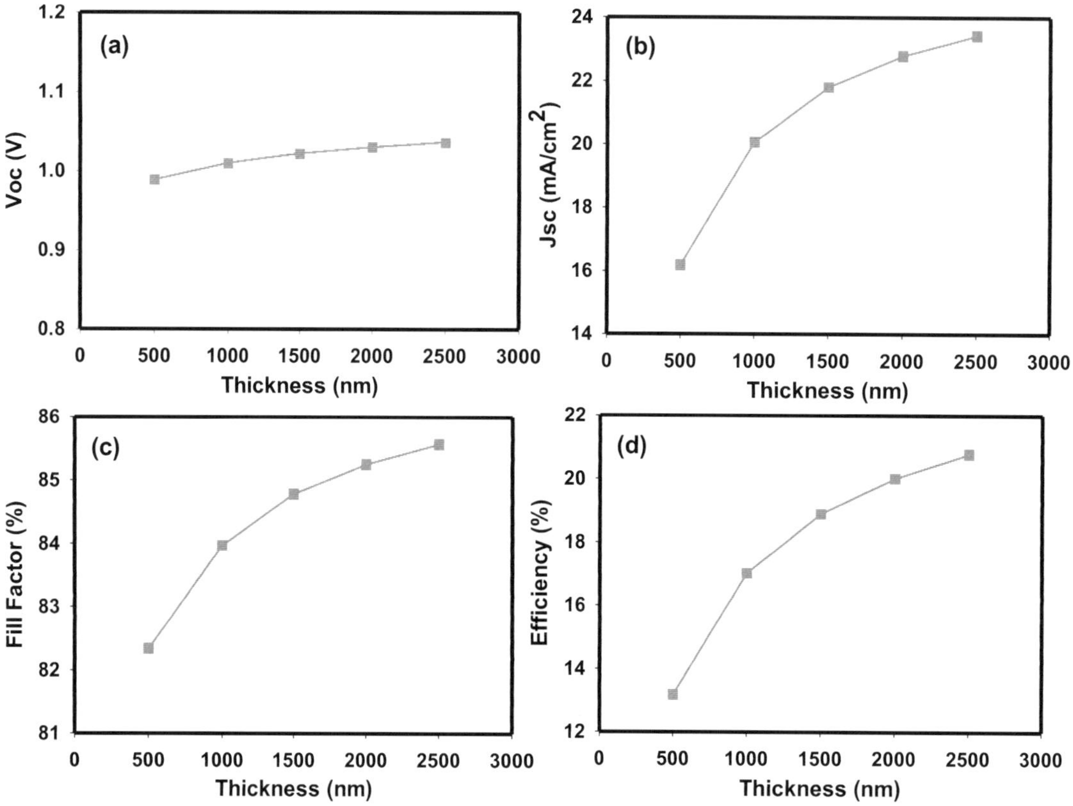

Figure 2. Impact of absorber layer thickness variations on the (a) V_{oc}, (b) J_{sc}, (c) FF and (d) efficiency.

3.2. Influence of CdTe carrier concentration

In this numerical analysis, the effect of CdTe carrier concentration is investigated. The carrier concentration is changed in the ranges of 1×10^{13} to 1×10^{17} cm^{-3} while other parameters remain constant. By varying the CdTe absorber layer carrier concentration, substantial variation in the cell performance has been observed. Figure 3 displays the effect of variable carrier concentration for CdS/CdTe structure while CdTe layer is fixed at thickness of 500 nm.

According to Figure 3, the combined effect of current density J_{sc} as well as the increase in V_{oc} and FF with the carrier concentration is responsible for the higher efficiency in the simulated results. It is obviously seen that the output parameters increase as the carrier concentration rises. The rise of carrier concentration resulted of increment in CdTe efficiency from 2.04% to 13.22%. As a result, carrier concentration of 1×10^{17} cm^{-3} gives the best performances for the ultrathin CdTe solar cells.

IOP Conf. Series: Materials Science and Engineering 1278 (2023) 012002 doi:10.1088/1757-899X/1278/1/012002

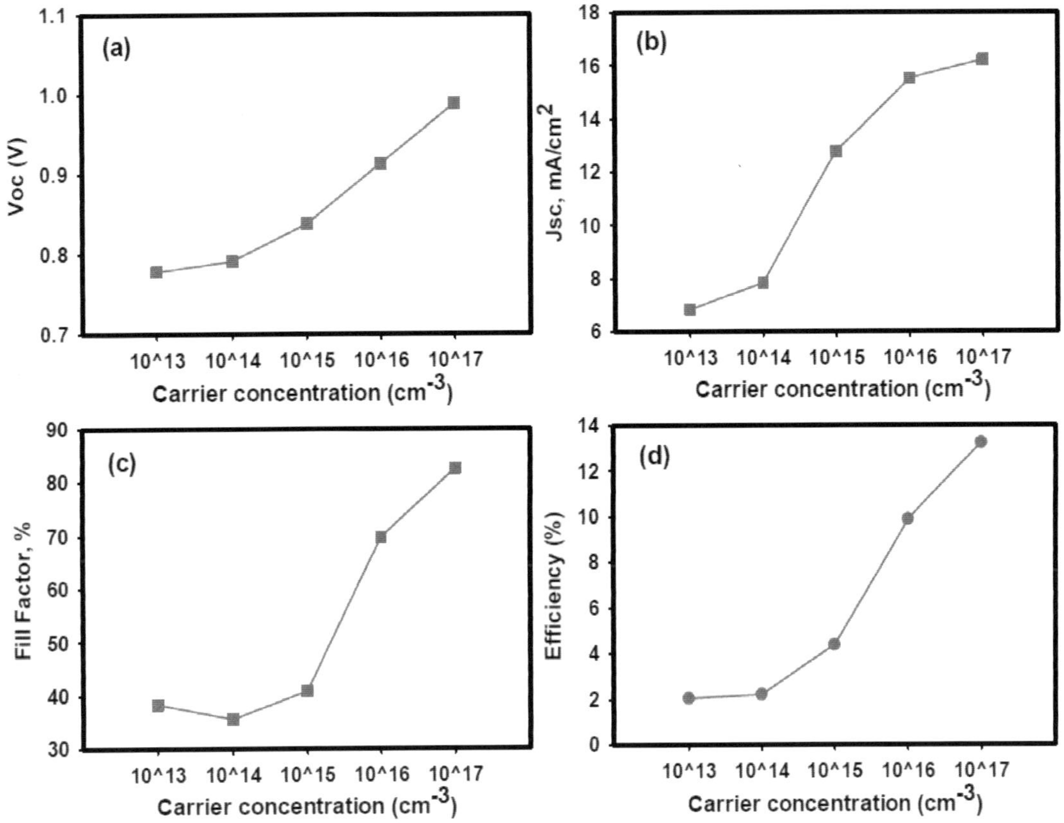

Figure 3. Influence of CdTe carrier concentration variations on (a) V_{oc}, (b) J_{sc}, (c) FF and (d) efficiency

3.3. J-V and QE

Quantum efficiency is a measure of the ratio of number of electrons collected out of cell at short circuit to the number of photon incident on cell. Measurements of quantum efficiency (QE) and current-voltage (J-V) provide important insights for understanding and improving the solar cell performance. The J-V characteristic with 500 nm thick CdTe layer are shown in Figure 4(a) which demonstrates the output characteristic of cell. Figure 4(b) signifies the light absorption is good for CdTe solar cell in the region from 300 – 900 nm. It is because of CdTe has a narrow band gap (1.5 eV) and poor response to solar shortwave spectrum.

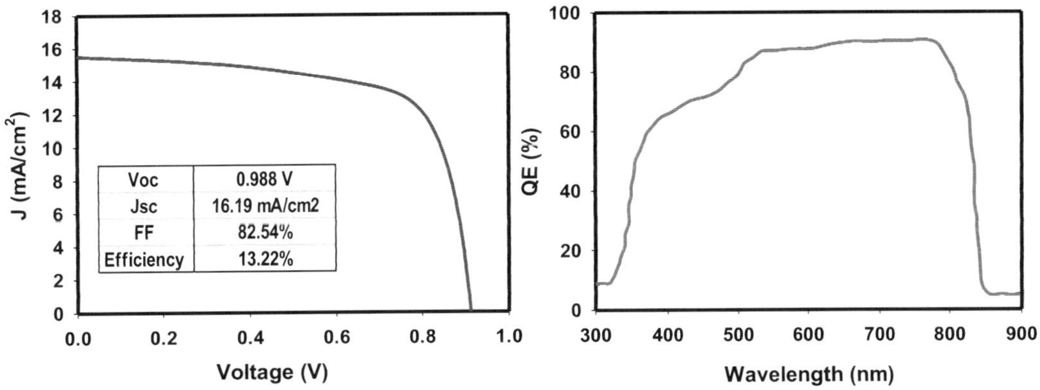

Figure 4. (a) J-V measurement and (b) Quantum Efficiency (QE) of CdS/CdTe solar cell structure.

3.4. *Effect of Operating Temperature*

The solar cell efficiency relies on the operating temperatures. The effects of temperature on CdTe cell performance have been examined using temperature variation from 300 K to 400 K, as shown in Figure 5. The simulation's output clearly shows that when the temperature rises, overall cell performance declines. Before reaching the depletion area and being collected, the electron has more energy and undergoes more recombination at higher temperatures.

Besides, a factor of the solar cell similar to electron or hole mobility, bandgap, and carrier concentration will influence the efficiency at high temperatures [3][12]. From Figure 5, all solar parameters decline as an increment in working temperature. The V_{oc} and J_{sc} output are dropped with the temperature rising from 300 K to 400 K due to open circuit potential being sensitive to the changes in temperature. A high temperature can affect the material conductivity, which can lead to degradation of CdTe solar cell performances and lead electron recombination happen before the depletion region [3]. The increased J_{sc} contributes to enhance the efficiency of solar cells.

Some factor that affects the fill factor are charge carrier recombination loss, the series resistance, reverse saturation current and shunt resistance. Due to the reverse saturation current's temperature dependence, V_{oc} decreases as temperature rises. The dependence of FF on the operating temperature can be deviated from the reliance of V_{oc} on the temperature. The relationship between FF and operating temperature is inversely proportional [13]. Temperature-related decreases in V_{oc}, J_{sc}, and FF result in temperature-related reductions in efficiency. The FF decreases sharply from 340 K to 400 K may be due to the increase in photo-generated carrier with illumination intensity which alter the resistance and influence the FF. Series resistance is related to intrinsic resistance of device and thickness of semiconducting layer. Meanwhile, shunt resistance is recombination or leakage current of device. Series resistance increases as the diffusion length shortens, reducing fill factor and efficiency [14].

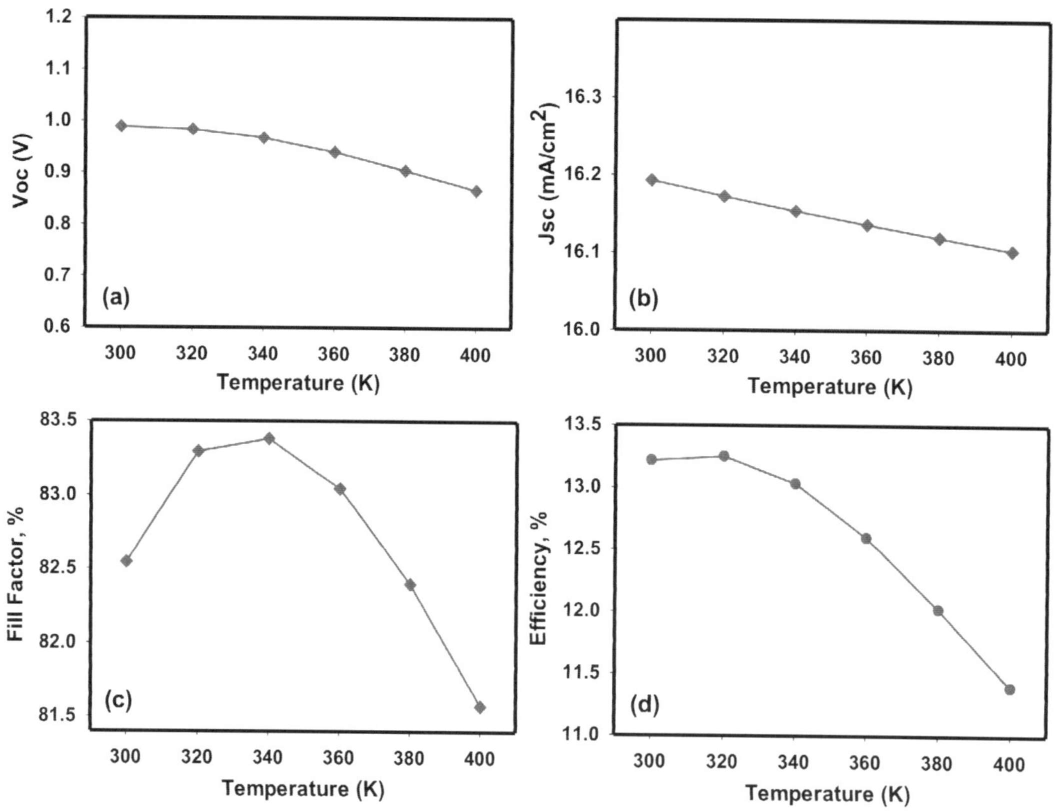

Figure 5. Impact of temperatures on (a) V_{oc}, (b) J_{sc}, (c) FF and (d) efficiency.

Figure 6 demonstrates the temperature gradient for CdS/CdTe structure. With operating temperatures ranging from 25°C to 125°C, an inquiry on normalized efficiency is conducted to determine the effects of high temperature on the output cell conversion efficiency. The effect of temperature is considered by the temperature coefficient, which quantifies the change in solar cell power output for an increase or decrease in operating temperature. The conversion efficiency evidently decreases linearly as operating temperature rises with temperature gradient of - 0.368/°C.

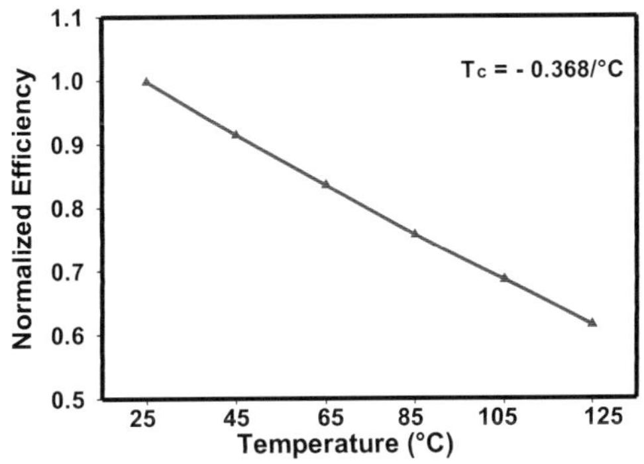

Figure 6. The temperature gradient for CdS/CdTe solar cell

Conclusion

In summary, the cell performance is simulated and analyzed by the function of the CdTe absorber layer thickness and carrier concentration by using the SCAPS-1D simulator. The production of ultrathin CdTe is suggested as a means of lowering the cost of CdTe solar cells. It is noticed that 500 nm thick CdTe absorber layer provides a reasonable efficiency over 10%. As the thickness of the absorber layer reduces, the J_{sc} and V_{oc} declines and accounts for the significant loss in the cell performance efficiency. The solar cell shows an efficiency of 13.22% (V_{oc} = 0.988 V, J_{sc} = 16.19 mA/cm^2, FF = 82.54%) with carrier concentration of 1.0×10^{17} cm^{-3} and 500 nm CdTe absorber layer. Moreover, the conversion efficiency decreases with the increase of operating temperature with a temperature coefficient of - 0.368/°C. As a result, it is expected that ultrathin CdTe solar cell can assist for an efficient and low-cost fabrication method.

Acknowledgement

The authors would like to acknowledge and appreciate the support of Centre for Research and Instrumentation Management (CRIM) of Universiti Kebangsaan Malaysia (UKM) through the research grant with the code of GGPM-2021-053. Due appreciation is also credited to the Ministry of Higher Education (MOHE), Malaysia for the funding through FRGS grant with the code of FRGS/1/2021/STG05/UKM/02/4. Additionally, the author would like to acknowledge Prof. Burgelman's group from University of Ghent, Belgium for providing the SCAPS-1D simulator.

References

[1] Doroody C, Rahman K S, Kiong T S, and Amin N, 2022 "Optoelectrical impact of alternative window layer composition in CdTe thin film solar cells performance," *Solar Energy* **233** 523–530.

[2] Isah M, Rahman K S, Doroody C, Harif M N, Rosly H N, Sopian K, Tiong S K and Amin N, 2021 "Design optimization of CdTe/Si tandem solar cell using different transparent conducting oxides as interconnecting layers," *Journal of Alloys and Compound.* **870** 159351.

[3] Enam F M T, Rahman K S, Kamaruzzaman M I, Sobayel K, Chelvanathan P, Bais B, Akhtaruzzaman M, Alamoud A R M, Amin N, 2017 "Design prospects of cadmium telluride/silicon (CdTe/Si) tandem solar cells from numerical simulation," *Optik (Stuttg).* **139** 397–406.

[4] Rashwan O, Sutton G, and Ji L, 2021 "Optical modeling of periodic nanostructures in ultra-thin CdTe solar cells with an electron reflector layer," *Superlattices Microstruct.* **149** 106757.

[5] Doroody C, Rahman K S, Rosly H N, Harif M N, Abdullah S F, Sopian K and N. Amin, 2021 "A comprehensive comparative study of CdTe thin films grown on ultra-thin glass substrates by close-spaced sublimation and RF magnetron sputtering," *Materials Letters* **293** 129655.

[6] Houshmand M, Zandi M H, and Gorji N E, 2015 "SCAPS Modeling for Degradation of Ultrathin CdTe Films: Materials Interdiffusion," *Jom*, **67**(9) 2062–2070.

[7] Huang C H and Chuang W J, 2015 "Dependence of performance parameters of CdTe solar cells on semiconductor properties studied by using SCAPS-1D," *Vacuum*, **118** 32–37.

[8] Doroody C, Rahman K S, Rosly H N, Harif M N, Haque F, Tiong S K and Amin N, 2020 "Impact of high resistivity transparent (HRT) layer in cadmium telluride solar cells from numerical simulation," *Journal Renewable Sustain. Energy*, vol. **12** 2.

[9] Biplab S R I, Ali M H, Moon M M A, Pervez M F, Rahman M F, and Hossain J, 2020 "Performance enhancement of CIGS-based solar cells by incorporating an ultrathin BaSi2 BSF layer," *J. Comput. Electron.* **19** (1) 342–352.

[10] Sarkar D K, Mahmud Hasan A K, Mottakin M, Vidhya S, Sobayel K, Ariful I M, Ghulam M, Aminuzzaman M, Shahiduzzaman M, Sopian K, Akhtaruzzaman M, 2022, Lead Free efficient perovskite solar cell device optimization and defect study using Mg doped $CuCro_2$ as HTL and WO_3 as ETL", *Solar Energy*, **243** 215-224.

[11] Ali M H, Moon M M A, and Rahman M F, 2019 "Study of ultra-thin CdTe/CdS heterostructure solar cell purveying open-circuit voltage ~1.2 v," *Material Research Express*, vol. **6** 9.

[12] Zahoo R K and Saleh A N, 2021 "Effect Of Carrier Concentration And Thickness Of Absorber Layer On Performance CBTS Solar Cell," *Turkish J. Comput. Math. Educ.* **12** (10) 5056–5064.

[13] Halim A M, 2022 "Simulation and Investigation of Cd-free SnS-based Solar Cell with a ZnSe as Buffer Layer using SCAPS-1D," *Journal of Circuits, System and Signal Processing*, vol **7** 10.

[14] Islam M T, Kumar A, 2021. Defect Density Control using an Intrinsic layer to enhance conversion Efficicicy in an Optimized SnS solar cell. *Journal of Electronic Materials.* **50** (3603-3613).

IOP Conf. Series: Materials Science and Engineering 1278 (2023) 012003 doi:10.1088/1757-899X/1278/1/012003

A numerical analysis of ZnTe/AZO as tunnel junction in CdTe/Si tandem solar cell

M Isah[1,4,6], C Doroody [2,6], K S Rahman[3], M N Harif [1,5], S K Tiong[2] and N Amin[2]

[1]College of Engineering, Universiti Tenaga Nasional (@The Energy University), Jalan IKRAM-UNITEN, 43000 Kajang, Selangor, Malaysia.
[2]Institute of Sustainable Energy, Universiti Tenaga Nasional (@The Energy University), Jalan IKRAM-UNITEN, 43000 Kajang, Selangor, Malaysia.
[3]Solar Energy Research Institute, Universiti Kebangsaan Malaysia, 43600 Bangi, Selangor, Malaysia.
[4]Department of Physics, Kaduna State University, P.M.B, 2339, Kaduna, Nigeria.
[5]Faculty of Applied Sciences, Universiti Teknologi MARA, Negeri Sembilan, 72000 Kuala Pilah, Malaysia.
[6]Author (s) to whom any correspondence should be addressed: almusty@gmail.com, camellia@uniten.edu.my.

ABSTRACT. Recently, interest has shifted towards developing multijunction or tandem solar cells due to their high potential to generate higher efficiency than traditional single-junction solar cells. Cadmium telluride (CdTe) and silicon (Si) solar cell materials have demonstrated significant potential in photovoltaic energy generation as tandem structures if fully developed. One approach for optimising CdTe/Si is to develop an effective tunnel junction that can electrically and optically interconnect the cadmium telluride and silicon cells with minimal loss. The wxAMPS 3.0 numerical simulation was used in this work to develop CdTe/Si tandem using zinc telluride/aluminium doped zinc oxide (ZnTe/AZO) as a tunnel junction (TJ). The result obtained shows an optimum efficiency of over 36 % with Voc = 1.945 V, Jsc = 21.519 mA/cm^2, and FF = 86.823 % utilising the optimal 200 nm CdTe and Si absorber thickness of 300 μm. An analysis of the demonstrated results suggests that ZnTe/AZO tunnel junction will significantly contribute to the realisation of the CdTe/Si tandem solar cell. Hence, upon inserting a 40 nm highly doped ZnTe/AZO tunnelling junction to a CdTe/Si tandem configuration, the solar cell's performance was enhanced by 48.190%.

Keywords: *Cadmium Telluride, Silicon, Tandem Solar Cell, ZnTe/AZO, Tunnel-Junction, wxAMPS, Numerical Simulations.*

1. Introduction

Tandem solar cells are currently receiving significant attention as they have shown immense potential in improving the efficiency of the present commercialised solar cells, which are mostly based on single junction technologies. Presently, two methods are commonly used to couple tandem solar cells for optimum performance, either mechanical stacking or monolithically. Any of the two methods will require the use of interconnecting layer or a tunnel junction (TJ) that is highly optical transparent and has a low electrical resistance which makes good ohmic contact with both the top and the bottom cells [1, 2]. However, one major bottleneck is selecting a suitable tunnel junction with high transparency and low current loss. In this work, a numerical study is carried out using ZnTe/AZO as a proposed tunnel junction, also referred to as tunnel diode for CdTe/Si tandem structure. ZnTe and AZO are selected in this study due to their high optical transparency and can make good ohmic contact with both CdTe and Si as a tunnel junction [3–6]. Zinc telluride is a binary chemical compound with a direct bandgap of 2.26 eV. It is normally a p-type semiconductor with a Zincblende structure (cubic). ZnTe has been used as a low-resistance back contact interlayer between CdTe and the metal contact in CdTe solar cells among other applications. The band offset of the interface between CdTe and ZnTe has been determined to be around 0.1 eV which is close to ideal for charge transport [7, 8]. A highly p-doped ZnTe can be obtained by incorporating Nitrogen (N) during its deposition [9]. Aluminium-doped zinc oxide (AZO) has a direct optical bandgap ranging between 3.46-3.54 eV depending on the aluminium contents, it is usually an n-type semiconductor compound. AZO is considered a good alternative transparent conducting oxide due to its good optoelectronic properties and low cost. It has many applications in optoelectronics devices because of its low resistivity and high optical conductivity [14]. Other combinations of materials such as AlGaAs/GaInP, AlGaAs/GaAs, AlGaAs/InGaAs, and AlGaAs/AlGaAs have been studied as tunnel diodes in different tandem solar cell structures [10–13]. CdTe/Si has been demonstrated to have the potential to ameliorate the power conversion efficiency from the present limit of either single junction CdTe or Si-based solar cells [15–18]. This is because CdTe/Si tandem solar cell combines the absorption of the spectral range for both CdTe and Si into a single solar cell due to their favourably different bandgaps (Si bandgap = 1.120 eV and CdTe bandgap = 1.500 eV) [19]. Silicon is the most established and commercialised solar cell material due to its earth abundance. The suitability of CdTe as a top absorber material for a two-junction (2J) tandem solar cell with Si has been demonstrated elsewhere [15, 16]. As a result of the considerable experimental resources needed to connect multijunction solar cells both electrically and optically require that optimisation is first carried out through device simulation [20, 21]. The proposed CdTe/Si tandem structure of the form SnO_2/CdS/p-CdTe/p^{++}-ZnTe/n^{++}-AZO/n-Si/p-Si/p+Si/Al was simulated using wxAMPS 3.0 numerical simulation software. In this study, the CdTe absorber thickness was optimised to match the current density of the coupled solar cells, the tunnel junction thickness was varied to strike a balance between the optical transparency and the current transport between the top and bottom cells, and the tunnel junction carrier concentration was also varied to obtain optimum value for the proposed device.

2. Simulation Method

In this work, wxAMPS 3.0 numerical simulation software is utilised. It is a one-dimensional, open-source software which is a modified version of AMPS-1D suited for simulating tandem solar cells [22]. It uses drift and diffusion models with features that include band-to-band tunnelling (BTBT), trap-assisted tunnelling (TAT), and intra-band tunnelling (IBT), which are used in modelling tandem solar cells. There are two types of band-to-band tunnelling models: local and non-local band-to-band tunnelling. The non-local band-to-band tunnelling is especially useful in simulating the tunnel junction used in multi-junction solar cells [23]. The non-local band-to-band tunnelling model has been incorporated in the wxAMPS 3.0 to effectively study the multi-junction devices [22]. Other features are bandgap narrowing analysis, field-dependent mobilities analysis, and transfer matrix method, as well as equipped with sub-cells analysis that enables the individual solar cells to be analysed and optimised separately [24]. A significant number of researchers have effectively used wxAMPS 3.0 software in their work [18, 22, 25–27]. In this work, this numerical simulation software is used to check the potential use and impact of p^{++}-ZnTe/p^{++}-AZO as a tunnel junction in a CdTe/Si tandem structure which has the form of SnO_2/n-CdS/p-CdTe/p^{++}-ZnTe/n^{++}-AZO/n-Si/p-Si/p+-Si/Al as shown in figure 1(b).

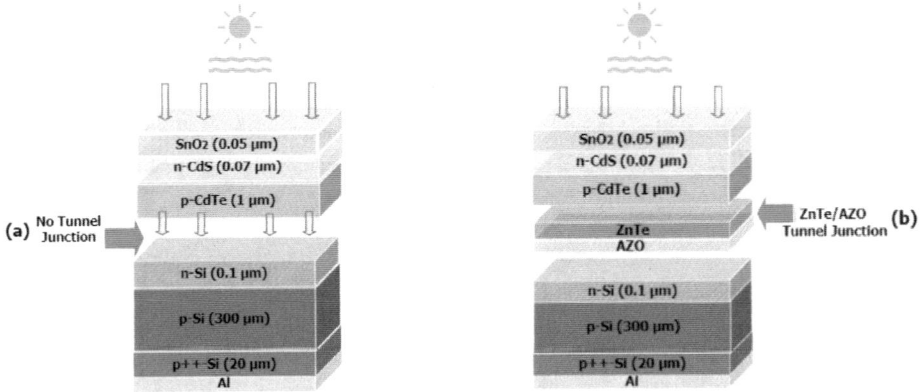

Figure 1. Schematic of CdTe/Si tandem solar cell (a) without and (b) with ZnTe/AZO tunnel junction.

The designed CdTe/Si tandem solar cell comprises a top cell based on CdTe solar cell structure (n-CdS/p-CdTe) with a Si bottom cell (n-Si/p-Si). The top and bottom sub-cells are first connected directly, as shown in figure 1(a) and simulated. Subsequently, the two sub-cells are connected using the p^{++}-ZnTe/n^{++}-AZO tunnel diode to see how they will impact the output of the designed CdTe/Si tandem device, as shown in figure 1(b). A tunnel junction or diode is designed to provide both low electrical resistance and optically low-loss connection between the sub-cells of a tandem solar cell. It also forms potential barriers for both the minority carriers generated by the top and bottom cells [28]. A tunnel junction (TJ), with a wide band gap and a highly doped diode, is usually used to connect the sub-cells of a tandem device. The wide band gap ensures no parasitic absorption of the remaining incident photons coming from the top cell for absorption by the bottom cell. High doping reduces the depletion region's length, making it easier for the generated carriers to tunnel through the depletion region as given, by equation (1).

$$l_{depl} = \sqrt{\frac{2\epsilon(\emptyset_0 - V)}{q} \frac{N_A + N_D}{N_A N_D}} \quad (1)$$

Where l_{depl} = depletion length, ϵ = dielectric permittivity, $\emptyset_0$ = built-in voltage, V = applied bias, q = electron charge, N_A and N_D are the amount of acceptor and donor carrier concentrations respectively.

Table 1. Simulation Parameters

Parameters	n-SnO_2	n-CdS	p++-ZnTe	n++-AZO	p-CdTe	n-Si	p-Si	p++-Si
Thickness (µm)	0.050	0.070	0.010-0.100	0.020	1.000 and 0.200	0.100	300	20.000
Electric Permittivity	9.000	10.000	9.670	10.000	9.400	11.900	11.900	11.900
Bandgap (eV)	3.400	2.420	2.260	3.400	1.500	1.120	1.120	1.120
Electron affinity (eV)	4.200	4.500	3.500	3.800	4.280	4.050	4.050	4.050
Conduction band effective density of States Nc (cm^{-3})	2.2×10^{18}	2.2×10^{18}	7.0×10^{16}	1.5×10^{18}	7.5×10^{17}	2.8×10^{19}	2.8×10^{19}	2.8×10^{19}
Valence band effective density of states Nv (cm^{-3})	1.8×10^{19}	1.8×10^{19}	2.0×10^{19}	1.8×10^{19}	1.8×10^{19}	1.04×10^{19}	1.04×10^{19}	1.04×10^{19}
Electron mobility Un (cm^2/Vs)	100	100	330	100	320	1350	1350	1350
Hole Mobility Up (cm^2/Vs)	25	25	80	50	40	450	450	450
Donor Concentration Nd	1×10^{17}	1×10^{17}	-	1×10^{18}	-	1.0×10^{16}	-	-
Acceptor Concentration Na	-	-	Variable	-	1.0×10^{14}	-	1.0×10^{16}	1.0×10^{19}

3. Results and discussion

3.1 The Baseline: CdTe/Si without the Tunnel Junction

To understand the impact of the proposed tunnel junction, a baseline CdTe/Si tandem device with a structure in the form of SnO_2/CdS/p-CdTe/n-Si/p-Si/p+Si/Al is shown in figure 1(a) is first simulated without the ZnTe/AZO tunnel junction. The output parameters obtained with this device, as demonstrated by the J-V curves in figure 2(a) are Voc = 1.242 V, Jsc = 13.358 mA/cm^2, FF = 86.387 % and an efficiency of 14.575% at a fixed CdTe absorber layer thickness of 1 μm and Si absorber layer thickness of 300 μm. To optimise the Jsc of this tandem structure, the optimum CdTe absorber layer thickness that gave the best efficiency was chosen to be 335 nm which offers an optimum efficiency of 17.772 % with Voc = 1.198 V, Jsc = 18.376 mA/cm^2 and a fill factor (FF) of 80.768 % as seen in figure 2(b). The poor output parameters obtained with this structure are clearly due to the direct connection between the top and bottom cells, making the p-type region of the top cell to be connected directly to the n-type region of the bottom cell as shown in figure 1(a), thereby forming a p-n junction having opposite polarities to the top and bottom sub-cells.

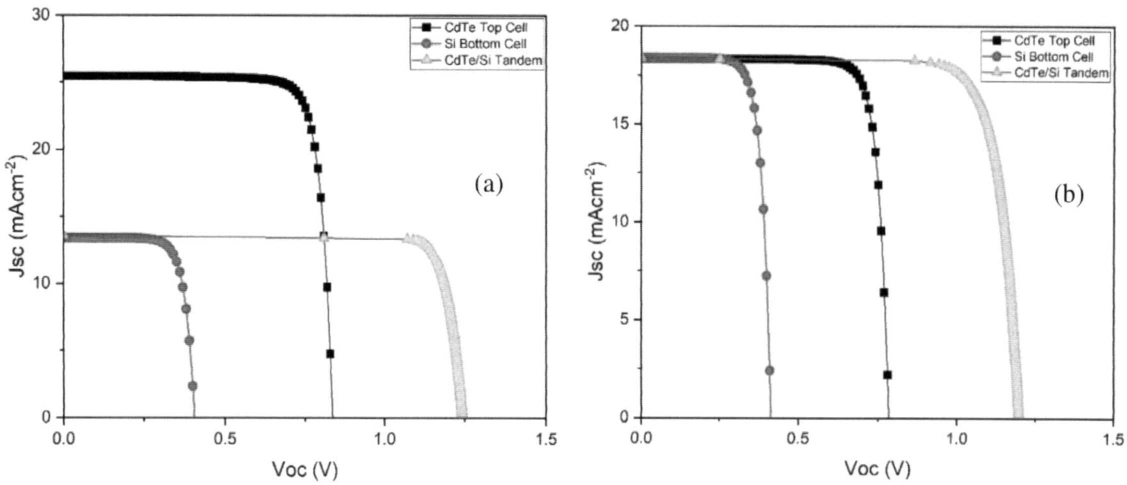

Figure 2. The J-V curve of the simulated (a) baseline CdTe/Si tandem cell and (b) the optimised CdTe/Si tandem cell without the TJ.

3.2 Impact of TJ on CdTe/Si tandem J-V Curve

A tandem device based on this structure SnO_2/CdS/p-CdTe/p^{++}-ZnTe/n^{++}-AZO/n-Si/p-Si/p$^+$Si/Al is simulated with wxAMPS 3.0 numerical simulation software to see how the proposed p^{++}-ZnTe/n^{++}-AZO tunnel junction will impact on CdTe/Si tandem solar cell parameters. As seen in figure 3, by fixing the CdTe absorber layer thickness at 1 μm, Si absorber layer at 300 μm, ZnTe, and AZO thickness were also fixed at 20 nm each. The initial output parameters obtained are Voc = 2.521 V, Jsc = 12.857, mA/cm^2, FF = 92.641 and efficiency (eff.) of 31.220 %. The small Jsc value obtained here may not be unconnected with the current of the top cell being constrained by that of the bottom cell. To optimise the thickness of the CdTe top cell to match the top and bottom cells' currents, the thickness of the CdTe absorber layer was set to 200 nm, which resulted in an improved Jsc to 21.519 mA/cm^2 from 12.857 mA/cm^2 and improved efficiency of 36.338 % from 31.220 %. These parameters are all represented by the J-V curves in figures 3(a) and b. Comparing the best efficiencies obtained between the baseline structure of 17.772 % and the one obtained from the proposed ZnTe/AZO tunnel junction of 36.338 % shows about 48.910 % improvement in the device performance. This may be attributed to the improved tunnelling current between the top and the bottom cells through the tunnel diode formed by highly doped ZnTe and AZO. This can be further corroborated by the band diagrams shown in figure 4. In Figure 4(a), the band diagram of the CdTe/Si tandem without any tunnel junction obtained from the simulation elaborates how some electrons will be collected in front, and others recombine at the back of the junction due to the absence of a tunnel junction. Hence, no contribution due to nonlocal band-to-band tunnelling, which is usually associated with monolithically connected tandem solar cells [23]. This hugely hinders the device's performance. Figure 4(b) shows how the conduction band of the Si cell is significantly narrowed towards the valence band of the CdTe top cell due to the highly doped ZnTe/AZO tunnel junction, which eases the tunnelling of electrons from the conduction band of Si to recombine with holes in the valence band, thereby generating tunnelling current between the subcells by non-local band-to-band tunnelling. This improves current collection and voltage addition of the top CdTe and bottom Si cells.

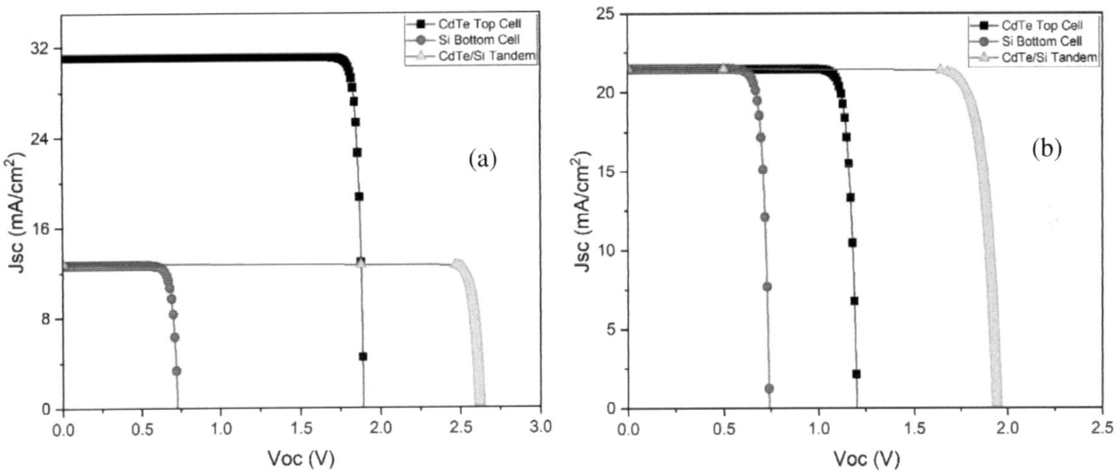

Figure 3. (a) J-V curve and (b) the optimised J-V curve of the CdTe/Si tandem cell with the ZnTe/AZO TJ.

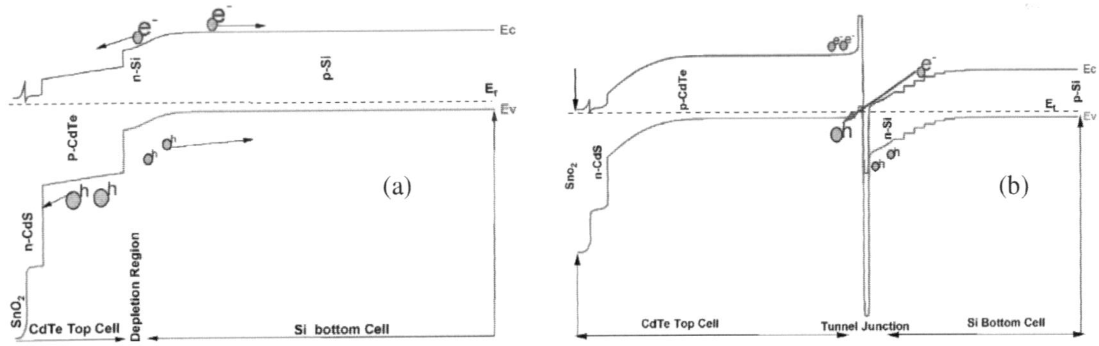

Figure 4. (a) Band diagram of CdTe/Si tandem solar cell (a) without and (b) with the heavily doped ZnTe/AZO tunnel junction.

3.3 Tunnel Junction thickness optimisation

Here, the thickness of the AZO was fixed at 20 nm, the CdTe absorber layer at 1 μm, and the Si absorber layer at 300 μm. The CdTe, ZnTe, and AZO carrier concentrations were all fixed, as given in table 1. The impact of ZnTe thickness is investigated by varying it from 10 nm – 100 nm. It can be seen that from figure 5, the Voc and FF of the device increased with the increasing thickness of the tunnel junction while the Jsc and efficiency decreased as the thickness of the tunnel junction increased. This can be due to more parasitic absorption of the sub-bandgap photons coming from the top cell for the bottom cell absorption by the ZnTe/AZO diode. The entire device response decreases with an increase in the tunnel junction thickness. An optimised thickness of 20 nm for the device is chosen in consideration of the experimental reality of achieving quality ZnTe thin film thickness.

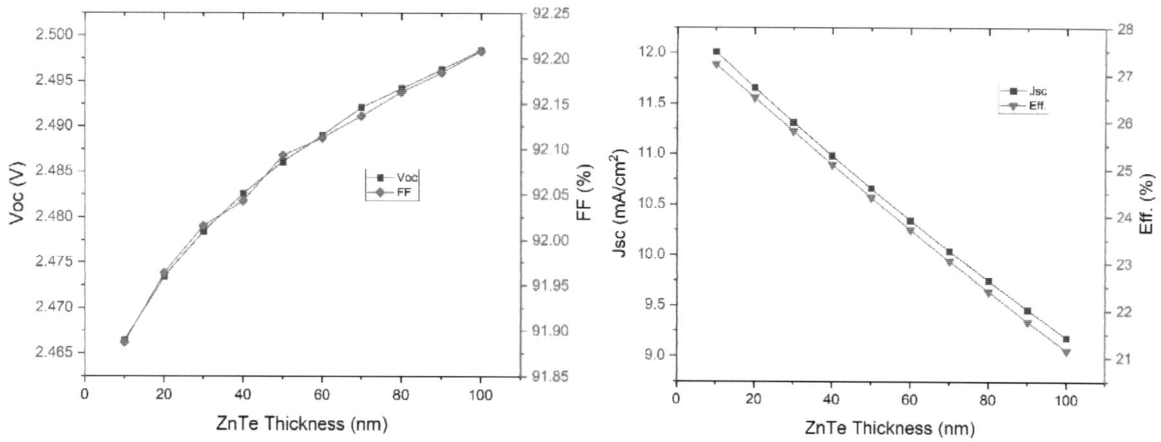

Figure 5. ZnTe thickness variation and the performance of the p^{++}-ZnTe/n^{++}-AZO tunnel junction.

3.4 Tunnel Junction Carrier Concentration optimization

By varying the carrier concentration for the ZnTe/AZO, the carrier concentration of AZO was first fixed at 1 x 10^{18} cm^{-2} while the carrier concentration of ZnTe was varied from 1 x 10^{10} to 1 x 10^{22} cm^{-22}, a considerable change in the cell efficiency is observed. Figure 6 shows the impact of this variation on the CdTe/Si tandem solar cell output parameters. It is seen that as the carrier concentration increases, the output parameters of the device also increase until the value of the carrier concentration becomes a little over 1 x 10^{22} cm^{-2} when the output parameters suddenly deteriorated probably as a result of an increase in interface recombinations with much higher value of carrier concentration. The initial increase in the output parameters of the device may be related to the depletion region becoming thinner to enable electrons to tunnel through it, this generates more currents across the tandem structure. Hence, forming a good ohmic contact with both the top and bottom cells.

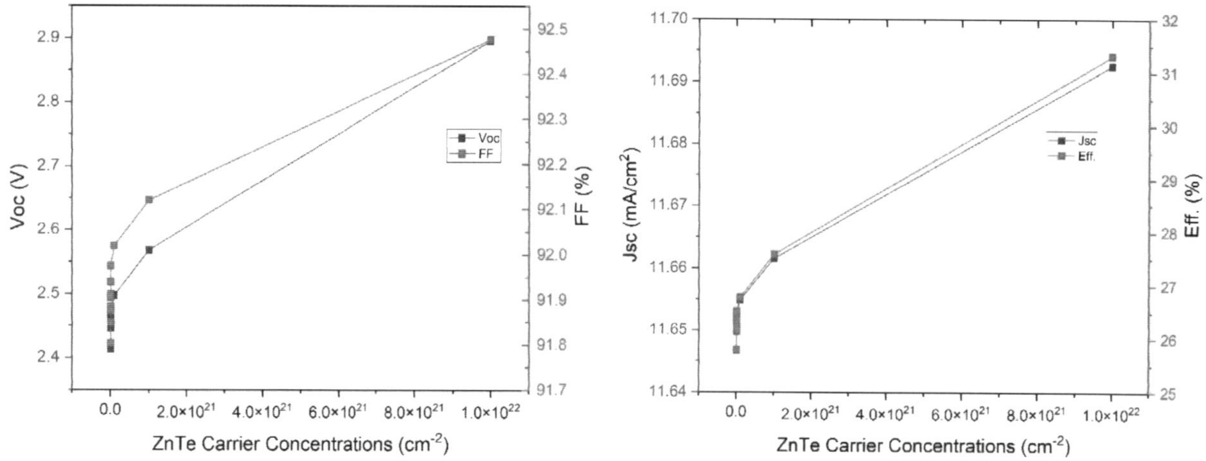

Figure 6. Impact of ZnTe carrier concentration variation on the performance of the p++ZnTe/n++AZO tunnel junction.

3.5 *Quantum efficiency*

Figure 7 shows how the increase in the tunnel junction thickness decreases the quantum efficiency of the bottom cell. This cannot be unconnected with the parasitic absorption of the photons coming from the top cell going to the bottom cell by the thicker tunnel diode. This reduces the overall performance of the tandem device, hence the optimum thicknesses of both ZnTe and AZO are chosen to be 20 nm.

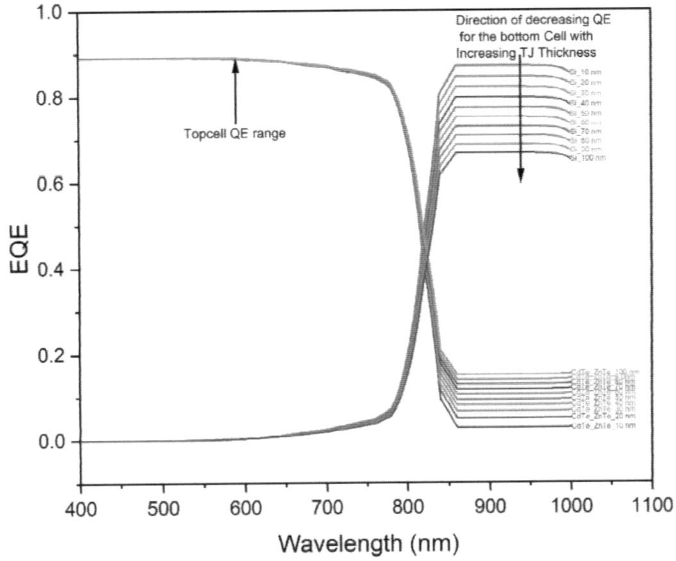

Figure 7. Impact of different tunnel junction thickness on the quantum efficiency of CdTe/Si device.

4. Conclusion

In this work, the impact of using ZnTe/AZO tunnel junction for CdTe/Si tandem solar cells has been studied using wxAMPS 3.0 simulation software. The impact of tunnel junction thickness and carrier concentrations have been investigated for the proposed structure. The results show over 48.91 % improvement in the overall device's performance by using the proposed ZnTe/AZO tunnel junction. Also, tunnel junction carrier concentration and thickness are demonstrated to have significant roles in designing a tunnel diode that can be used to effectively connect tandem devices both optically and electrically. The overall result indicates that the proposed p^{++}-ZnTe/n^{++}-AZO tunnel diode can be used to fabricate an effective and more efficient CdTe/Si tandem solar cell. This, if achieved, will greatly contribute to realising a more robust CdTe/Si two-junction solar cell that can compete with other emerging multi-junction devices in extending the theoretical limit of the single-junction solar cell.

Acknowledgement

The authors would like to acknowledge the Ministry of Higher Education of Malaysia (MOHE) for the support given by the HICoE grant for this research with the code 2022003HICOE parked at the iRMC of Universiti Tenaga Nasional (UNITEN). We would also like to acknowledge Prof. Rockett, Dr. Yiming Liu of UIUC and Prof. Fonash of PSU for developing the wxAMPS 3.0 numerical software that we used in this work.

References

[1] M. Baudrit and C. Algora, 'Tunnel diode modeling, including nonlocal trap-assisted tunneling: A focus on IIIV multijunction solar cell simulation', *IEEE Trans Electron Devices*, vol. 57, no. 10, pp. 2564–2571, 2010, doi: 10.1109/TED.2010.2061771.

[2] N. Moulin, M. Amara, F. Mandorlo, and M. Lemiti, 'Tunnel junction i (V) characteristics: Review and a new model for p-n homojunctions', *J Appl Phys*, vol. 126, no. 3, 2019, doi: 10.1063/1.5104314.

[3] A. Bosio *et al.*, 'Interface phenomena between CdTe and ZnTe:Cu back contact', *Solar Energy*, vol. 176, pp. 186–193, Dec. 2018, doi: 10.1016/j.solener.2018.10.035.

[4] M. Boumaour, S. Sali, S. Kermadi, L. Zougar, A. Bahfir, and Z. Chaieb, 'High efficiency silicon solar cells with back ZnTe layer hosting IPV effect: a numerical case study', *Journal of Taibah University for Science*, vol. 13, no. 1, pp. 696–703, 2019, doi: 10.1080/16583655.2019.1623476.

[5] K. H. Kong, R. Kek, T. Y. Tou, and S. S. Yap, 'Effects of the resistivity of AZO film on the IV and CV characteristics of AZO/p-Si heterojunction', *Microelectron Eng*, vol. 213, no. March, pp. 24–30, 2019, doi: 10.1016/j.mee.2019.04.011.

[6] H. Bo *et al.*, 'Characterization of AZO/p-Si heterojunction prepared by DC magnetron sputtering', *Mater Sci Semicond Process*, vol. 12, no. 6, pp. 248–252, 2009, doi: 10.1016/j.mssp.2009.12.006.

[7] B. Späth, J. Fritsche, F. Säuberlich, A. Klein, and W. Jaegermann, 'Studies of sputtered ZnTe films as interlayer for the CdTe thin film solar cell', in *Thin Solid Films*, Jun. 2005, vol. 480–481, pp. 204–207. doi: 10.1016/j.tsf.2004.11.073.

[8] C. Hernandez-Vasquez, M. A. González-Trujillo, S. Gallardo-Hernández, and M. L. Albor-Aguilera, 'Unveiling the influence of ZnTe and Te layers as part of the back-contact on CdTe solar cells performance', *AIP Adv*, vol. 11, no. 3, Mar. 2021, doi: 10.1063/5.0043595.

[9] T. Baron, K. Saminadayar, and N. Magnea, 'Nitrogen doping of Te-based II-VI compounds during growth by molecular beam epitaxy', *J Appl Phys*, vol. 83, no. 3, pp. 1354–1370, Feb. 1998, doi: 10.1063/1.366838.

[10] B. Wang *et al.*, 'Effects of Zn diffusion in tunnel junction and its solution for high efficiency large area flexible GaInP/GaAs/InGaAs tandem solar cell', *Solar Energy Materials and Solar Cells*, vol. 230, p. 111257, Sep. 2021, doi: 10.1016/j.solmat.2021.111257.

[11] Y. Sun *et al.*, 'Delta-Doping for Enhanced III-V Tunnel Junction Performance', *IEEE J Photovolt*, vol. 12, no. 4, pp. 976–981, Jul. 2022, doi: 10.1109/JPHOTOV.2022.3176217.

[12] Y. Gou, H. Wang, J. Wang, H. Yang, and G. Deng, 'High performance p++-AlGaAs/n++-InGaP tunnel junctions for ultra-high concentration photovoltaics', *Opt Express*, vol. 30, no. 13, p. 23763, Jun. 2022, doi: 10.1364/oe.461302.

[13] M. Yamaguchi, T. Takamoto, A. Khan, M. Imaizumi, S. Matsuda, and N. J. Ekins-Daukes, 'Super-high-efficiency multi-junction solar cells', in *Progress in Photovoltaics: Research and Applications*, Mar. 2005, vol. 13, no. 2, pp. 125–132. doi: 10.1002/pip.606.

[14] K. N. Tonny, R. Rafique, A. Sharmin, M. S. Bashar, and Z. H. Mahmood, 'Electrical, optical and structural properties of transparent conducting Al doped ZnO (AZO) deposited by sol-gel spin coating', *AIP Adv*, vol. 8, no. 6, Jun. 2018, doi: 10.1063/1.5023020.

[15] F. M. T. Enam *et al.*, 'Design prospects of cadmium telluride/silicon (CdTe/Si) tandem solar cells from numerical simulation', *Optik (Stuttg)*, vol. 139, pp. 397–406, 2017, doi: 10.1016/j.ijleo.2017.03.106.

[16] A. C. Tamboli, D. C. Bobela, A. Kanevce, T. Remo, K. Alberi, and M. Woodhouse, 'Low-Cost CdTe / Silicon Tandem Solar Cells', vol. 7, no. 6, pp. 1767–1772, 2017.

[17] M. Carmody *et al.*, 'Single-crystal II-VI on Si single-junction and tandem solar cells', *Appl Phys Lett*, vol. 96, no. 15, 2010, doi: 10.1063/1.3386529.

[18] M. Isah *et al.*, 'Design optimization of CdTe/Si tandem solar cell using different transparent conducting oxides as interconnecting layers', *J Alloys Compd*, vol. 870, Jul. 2021, doi: 10.1016/j.jallcom.2021.159351.

[19] B. Zhou *et al.*, 'Numerical simulation of an innovative high efficiency solar cell with CdTe/Si composite absorption layer', *Opt Mater (Amst)*, vol. 110, no. September, p. 110505, Dec. 2020, doi: 10.1016/j.optmat.2020.110505.

[20] Y. G. Xiao, Z. Q. Li, M. Lestrade, and Z. M. S. Li, 'Modeling of CdZnTe/CdTe/Si triple junction solar cells', *Conference Record of the IEEE Photovoltaic Specialists Conference*, no. 604, pp. 001386–001389, 2011, doi: 10.1109/PVSC.2011.6186215.

[21] A. Amine, Y. Mir, and M. Zazoui, 'Modelling of dual-junction solar cells including tunnel junction', *Advances in Condensed Matter Physics*, vol. 2013, 2013, doi: 10.1155/2013/546362.

[22] Y. Liu, Y. Sun, and A. Rockett, 'A new simulation software of solar cells - WxAMPS', *Solar Energy Materials and Solar Cells*, vol. 98, pp. 124–128, 2012, doi: 10.1016/j.solmat.2011.10.010.

[23] M. Hermle, G. Létay, S. P. Philipps, and A. W. Bett, 'Numerical simulation of tunnel diodes for multi-junction solar cells', *Progress in Photovoltaics: Research and Applications*, vol. 16, no. 5, pp. 409–418, Aug. 2008, doi: 10.1002/pip.824.

[24] Y. Liu, M. Ahmadpour, J. Adam, J. Kjelstrup-Hansen, H. G. Rubahn, and M. Madsen, 'Modeling Multijunction Solar Cells by Nonlocal Tunneling and Subcell Analysis', *IEEE J Photovolt*, vol. 8, no. 5, pp. 1363–1369, 2018, doi: 10.1109/JPHOTOV.2018.2851308.

[25] W. Chen, X. Huang, Q. Cheng, C. Chen, D. Yun, and F. Zhang, 'Simulation analysis of heterojunction ZnO/CdS/Cu(In,Ga)Se2 thin-film solar cells using wxAMPS', *Optik (Stuttg)*, vol. 127, no. 1, pp. 182–187, 2016, doi: 10.1016/j.ijleo.2015.09.198.

[26] T. Gokmen, O. Gunawan, and D. B. Mitzi, 'Semi-empirical device model for Cu2ZnSn(S,Se)4 solar cells', *Appl Phys Lett*, vol. 105, no. 3, pp. 2–7, 2014, doi: 10.1063/1.4890844.

[27] Y. Wang *et al.*, 'Towards printed perovskite solar cells with cuprous oxide hole transporting layers: A theoretical design', *Semicond Sci Technol*, vol. 30, no. 5, p. 54004, 2015, doi: 10.1088/0268-1242/30/5/054004.

[28] M. Yamaguchi, T. Takamoto, and K. Araki, 'Super high-efficiency multi-junction and concentrator solar cells', *Solar Energy Materials and Solar Cells*, vol. 90, no. 18–19, pp. 3068–3077, Nov. 2006, doi: 10.1016/j.solmat.2006.06.028.

Design and Simulation of Lead-free Flexible Perovskite Solar cell Using *SCAPS-1D*

A A Goje[1,2], N A Ludin[1], M A Mat Teridi[1], U Syafiq[1], M A Ibrahim[1], F Nawab[1] and A A Syakirin[1]

[1]Solar Energy Research Institute (SERI), Universiti Kebangsaan Malaysia, 43600, Bangi, Selangor Darul Ehsan, Malaysia.

[2]Department of Science Laboratory Technology, Federal Polytechnic Damaturu, Yobe State Nigeria.
E-mail: sheekeen@ukm.edu.my

Abstract. Despite perovskite solar cells' rapidly increasing efficiency, industrializing and commercializing the device presents several challenges. Top among these is toxicity due to the material's lead-based perovskite usage, which causes rapid degradation. Because of this challenge, researchers have recently become interested in lead-free perovskite materials. In this research, we investigate using the SCAPS simulation software to optimize a lead-free flexible solar cell. Lead (Pb), which belongs to group 14, can likely be replaced with bismuth (Bi), which belongs to group 15. In this study, the structure of the device is an n-i-p planar lead-free heterostructure flexible perovskite solar cell (FPSC) comprised of a Polyethylene Terephthalate/Indium Tin Oxide (PET/ITO) substrate; [6,6]-phenyl C61 butyric acid methyl ester (PCBM) Electron Transport Layer (ETL); methyl ammonium bismuth iodide ($CH_3NH_3BiI_3$) as an absorber; Spiro-OMeTAD as a Hole Transport Layer (HTL); and silver (Ag) as the electrode. We optimised cell performance and efficiency with an absorber layer of 100 nm and an optimised temperature of 300 K. This absorber layer's thickness and defect density were considered. The defect density of the FPSC is indirectly proportional to the device's absorption coefficient (alpha). We obtained a power conversion efficiency (PCE) of 18.80%, a current density (J_{SC}) of 35.59 mA/cm^2, an open circuit voltage (V_{OC}) of 0.633 V, and a fill factor (FF) of 83.43%. We also observed that the FF is inversely proportional to the thickness of the absorber layer of the FPSC.

.

Keywords: Flexible perovskite solar cell, lead-free perovskite, Bismuth, SCAPS simulation software.

1. Introduction

Perovskite materials are now being considered in various optoelectronic applications due to their superior optoelectronic characteristics and ease of manufacturing [1] [2]. FPSC has increased rapidly in PCE from 3.8% to 24.7%, which indicates they have great potential for photovoltaic applications [3]. Demand for FPSCs is developing due to the expanding market for self-powered flexible and lightweight wearable electronics. Its perovskite film offers advantages in fabricating flexible devices because of its flexibility and low processing temperature. Additionally, FPSCs benefit from cheap costs due to the roll-to-roll (R2R) fabrication technique; this makes the fabricating process faster and enables the production of large-scale devices. Another advantage of the FPSCs is that they are cheaper to transport and install because they are lightweight. Moreover, FPSCs outperform other flexible solar cells (FSC), including cadmium telluride (CdTe) and copper indium gallium selenide (CIGS), in terms of power-per-weight, which is good for the use of wearable and lightweight electronics devices [4]. FPSCs have several additional advantages, which include self-healing, biocompatibility, stretchability, good efficiency, and environmental friendliness [5].

However, toxicity, air instability, moisture, and high temperatures are some of the challenges that perovskite solar cells (PSC) must overcome to be accepted for commercial purposes, despite extensive research in the field [6]. To improve effectiveness and reduce the bandgaps and other performances, including diffusion length, several non-toxic materials have been investigated. Early attempts to address the toxicity issue included replacing lead with tin (Sn) in $CsSnI_3$, but the effectiveness was limited to just 0.9% [7]. Other attempts have been made to fabricate lead free PSCs and improve their performance. For instance, L. Fu et al. used bismuth telluride (Bi_2Te_3) to replace Pb in the HTL/perovskite interface structure and increased the device PCE by 11.9%; the findings demonstrate that IL prevents iodide ions from migrating from the absorber layer [8].

Furthermore, because of its non-toxicity, high absorption coefficient, narrow bandgap (1.72 eV), and short carrier lifetimes, bismuth iodide (BiI_3) material has been receiving increasing attention. Because of these characteristics, BiI_3 may be more suitable for solar applications [9]. Recently, Y. Hu et al. fabricated a PSC device with pure BiI_3 over the titanium oxide ETL and achieved an enhanced electron transport behaviour by passivating the interfacial trap states. They obtained a PCE of 17.8% [9]. Figure 1 below demonstrates that a strong anti-bonding interaction between the Pb-s and I-p orbitals accounts for most of the $CH_3NH_3PbI_3$ upper valence band dispersion and that of $CH_3NH_3BiI_3$. The high diffusion length for electron-hole pairs, which is unusual for PSC applications, is made possible by the strong s-p anti-bonding interaction, bringing the effective mass of holes close to that of electrons [10].

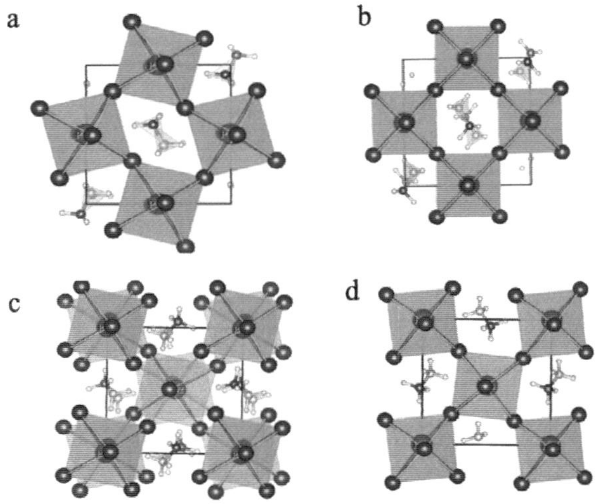

Figure 1. The orbital structure for O-$CH_3NH_3PbI_3$ in (010) plane (a) O-$CH_3NH_3BiI_3$ in (010) plane (b), T-$CH_3NH_3PbI_3$ in (001) plane (c), and (d) T-$CH_3NH_3BiI_3$ in (001) plane. [10]

Over the years, SCAPS-1D has been commonly used in research on solar cells, including PSC, to explore the effects of the device design and characteristics on their performance and to simulate the thickness of layered structures and other parameters. Using SCAPS-1D, Kai Tan and colleagues built solid-state PSCs. They reported that due to the good bandgap, chemical interaction, and large conductivity found with a perovskite absorber, the selected HTL effect on the performance of PSC with

IOP Conf. Series: Materials Science and Engineering 1278 (2023) 012004 doi:10.1088/1757-899X/1278/1/012004

copper conductors (CuI) and thiophene polymer (PTAA) achieved higher PCE [11]. Lin et al. utilised SCAPS to design the HTL-free PSC and obtained PCE of 15% at normal conditions [12]. In the simulation of PSC at different charge carrier mobilities, Narender et al. designed PSC with an efficiency of 20% at hole mobilities of 0.510 x 10^{-6} $m^2v^{-1}s^{-1}$[13]. Usha et al. simulated an organic-inorganic PSC and examined the impact of active layer thickness and defect density on the efficiency of the PSC. They obtained a J_{sc} of 25.60 $mAcm^{-2}$ and a PCE of 31.77% [14]. To develop acceptable FPSCs while avoiding lead toxicity, new types of PSC must be considered an alternative. The key motivation is to examine replacing lead with an element closer to it in the periodic table to preserve lead's special features and prevent toxicity; bismuth (Bi) is a suitable candidate. An analysis of its electrical, optical, and transport properties was conducted to determine whether it satisfies the requirements for being a good absorber layer for photovoltaic solar cells [8]. The ability of $CH_3NH_3PbI_3$ to absorb light in the infrared and visible regions is one benefit of choosing bismuth. Other benefits include the element's abundance and the fact that it is non-toxic and environmentally benign. We also considered avoiding the indirect bandgap. However, though silicon, the most common absorber layer, has an indirect bandgap, doing so requires a thicker layer, which can be troublesome in low mobilities. As a result, we will discuss an FPSC analysis of photovoltaic systems utilizing completely lead-free materials and methylammonium bismuth iodide ($CH_3NH_3BiI_3$) as the absorber material. The excellent performance of the hybrid organic-inorganic metal halide PSCs is comparable to silicon-based solar cells. They have a tunable bandgap, a lower excitation binding energy, high optoelectrical properties, and a longer carrier diffusion distance. This article uses a 1D-solar cell capacitance simulator to replicate the device's operation in AM1.5G light (SCAPS, ver.3.3.07). The doping concentrations and absorber layer thickness are considered for optimizing performance. However, no Bi replaced lead free-FPSC-based simulation, as theoretical analysis has yet been documented in the PSC study; as a result, we will investigate it in this project utilizing the software.

2. Device structures

The absorption layer, $MABiI_3$, is assumed to be deposited between the ETL/HTL layers in the device architecture, as shown in figure 2 below. The ETL, HTL, and absorber layer thicknesses are varied over a broad range to enhance PCE. As a result, the device simulation of an FPSC based on $MABiI_3$ with a planar heterojunction of PET/ITO/PCBM/$MABiI_3$/Spiro-OMeTAD/Ag was simulated through n-i-p mode. Silver (Ag) was selected as the metal back contact, and light incidences across PET/ITO under the work function of 4.8 eV, were used as ITO and considered to have a thickness of 300 nm with a transmissivity of 89% and a surface resistance of 15 ohm/sq. The simulation uses the solar cell capacitance simulator under the 1 Sun spectrum of the software simulator. The modelling process accounts for the thicknesses of the $CH_3NH_3BiI_3$ perovskite photoactive layer and considers the defect and the effect of thickness on the FF.

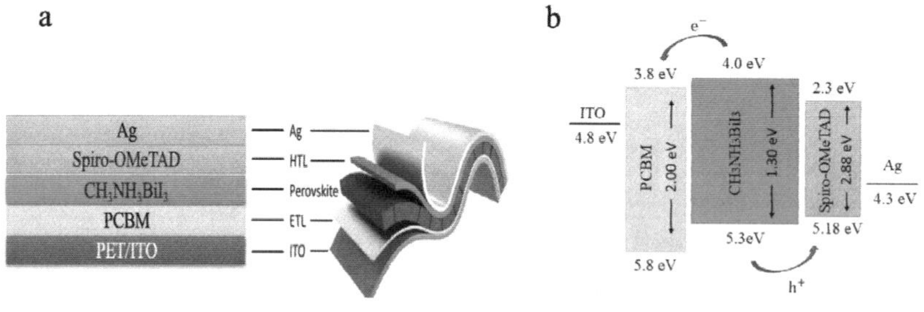

Figure 2. (a) The architecture of the FPSC devices simulated (b) the energy band diagram

3. Simulation Analysis and Methodology

A simulation method is used to numerically model FPSCs under the 1-sun spectral irradiance to affect the spectral response of solar energy with different homo-junction and heterojunction architectures [15]. The process is performed numerically based on a solution using Poisson's and semiconductor equations to specify the physics of the photovoltaic structure and tracking behavior necessary for charge carrier mobility. A continuity equation in the position domain, which determines the solar cell's electrostatic potential, depicts the transport of charge carriers inside the device, including electron-hole recombination and Poisson's equation [16]. The equation is given by:

$$\nabla^2\Psi = \varrho\varepsilon\,(n - p + N_A - N_D) \qquad (1)$$

N_A is the acceptor doping, while donor doping (N_D), ε stands for material permittivity and Ψ represent electrostatic potential, and the semi-conductor equation is given as;

$$\nabla.J_n - \varrho\frac{\partial n}{\partial t} = +qR. \qquad (2)$$

$$\nabla.J_p - \varrho\frac{\partial n}{\partial t} = -qR. \qquad (3)$$

J_n is the current density for electrons conducted for current flow, and to transfer the charge, E electric field, ϱ is the fundamental charge unit [15] continuity equation is given by

$$J_n = \varrho n\mu_n E + \varrho D_n\nabla_n \qquad (4)$$

$$\nabla J_p = \varrho n\mu_p E + \varrho D_p\nabla_{p.} \qquad (5)$$

Then from the continuity equation--.

The diffusion coefficients for holes and electrons are denoted by the D_p and D_n, respectively. The software interfaces utilise all variables listed in table 1 and are the device's input parameters during the simulation. This demonstrates that the output of the device simulation is accurate and that the input parameters mimic that of the physical device. Additionally, it displays a list of the parameters taken into consideration through the procedure to achieve the optimise result by changing the variables and investigating how various elements would impact a device's performance. A detail of the simulation procedure is shown in figure 3 below.

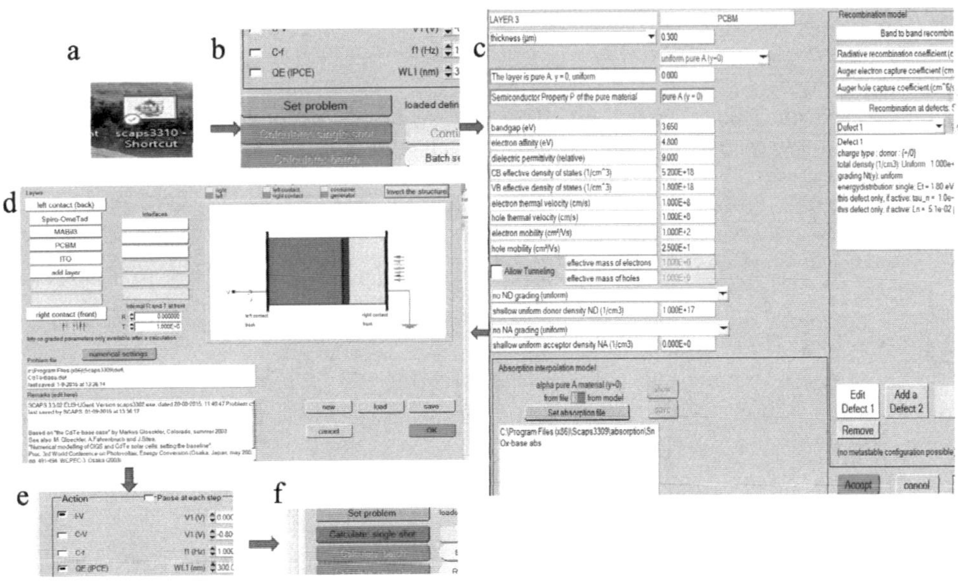

Figure 3. The SCAPS 1D simulation procedure (a) launching the SCAPS application software (b) setting a problem (c) inputting parameters (d) setting working conditions (e) specifying the actions to be measured (d) running the program

3.1 Thickness Optimization

Layer thickness simulation was performed using simulation software. Before SCAPS 3.3.00, which was released in 2014, layer thickness input values were expressed in micrometers (µm), with a thickness range of 1 nm (0.001 µm) to 10 cm (10^5 µm) and a display in three decimal digits. The input of a thickness lower than 1 nm was automatically set to 0.001 µm = 1^{-5} nm, which was tedious when one had input d = 0.0025 µm (thus 2.5 nm): the software would calculate 2.5 nm, and show 0.002 m (thus 2 nm). The permitted input range of thickness is 0.01 nm = 10^{-5} µm. Also, 1 m = 10^6 µm. The user can select from various units for the thickness display, Å (1 Ångström = 0.1 nm), nm, µm, mm, and cm. However, the number of decimal digits is limited to three. When a measurement other than the conventional micrometer is chosen, the value and measurement are shown in magenta [17]. . By setting the display unit, to µm, for example, we need to input 0.00245, and the display will show the result rounded to three digits, or 0.002 µm; however, when the display unit is set to nm, 2.45 nm will be displayed (see Figure 4). The thickness will be shown in nm if d 10 nm, nm if d 1000 m, and mm if d > 1 mm whenever a newer problem is imported from a file. We did this by altering the thickness of the absorber layer, which is the $MABiI_3$ perovskite layer.

a b

Figure 4. The layer thickness display with d = 0.00245 µm input and m as the display unit in black (a), and with nm as a display unit in pink colour (b) [17]

4. Result and Discussion

Measurements of structural profilometry and experimental absorption spectra are used to determine the Spiro-OMeTAD (HTL) thickness (350 nm [20]. The simulation also used the experimental PCBM (ETL) thickness (30 nm) [22]. From the literature, other variables were taken, including electron affinity, dielectric permittivity, CB effective density of states, VB effective density of states, charge mobility, thermal velocities, doping densities, and defect densities [18] [19], [21], [23]. The V_{oc} against Jsc for different photoactive layer thicknesses is illustrated in figure 4 below. These values were determined using the simulation software, and the results were plotted using the Origin 2021 software. We found that the Jsc decreases marginally with a thickness of 30 nm and 300 nm once the $CH_3NH_3BiI_3$ perovskite layer has a 100 nm thickness. As a result, we consider that our FPSC should have a 100 nm optimum thickness.

Table 1: Materials properties of various layers simulated

Parameters	ITO	PCBM	$CH_3NH_3BiI_3$	Spiro-OMeTAD
Thickness (nm)	300	30	Varied	350
Band gap (eV)	3.65	2.00	1.3	2.88
Electron affinity (eV)	4.80	3.90	4.17	2.05
Dielectric permittivity	8.90	4.00	6.5	3.00
CB effective density of states ($1/cm^3$)	5.2×10^{18}	1×10^{12}	1×10^{18}	2.2×10^{18}
VB effective density of states ($1/cm^3$)	1.0×10^{18}	2×10^{20}	1×10^{19}	1.8×10^{19}
Electron thermal velocity (cm/s)	1.0×10^8	1×10^7	1×10^7	1×10^7
Hole thermal velocity (cm/s)	1.0×10^8	1×10^7	1×10^7	1×10^7
Electron mobility (cm^2/ V.s)	1.0×10^{12}	1×10^{-2}	1.6	2.0×10^{-4}
Hole mobility (cm^2/ V.s)	1.0×10^{12}	1×10^{-2}	1.6	2.0×10^{-4}
Shallow uniform donor density N_D ($1/cm^3$)	1.0×10^{12}	1×10^{20}	0	0
Shallow uniform acceptor density N_A ($1/cm^3$)	0	0	3.2×10^{13}	2.0×10^{19}
Defect density Nt ($1/cm^3$)	1×10^{14}	1×10^{14}	1.0×10^{14}	1.0×10^{15}

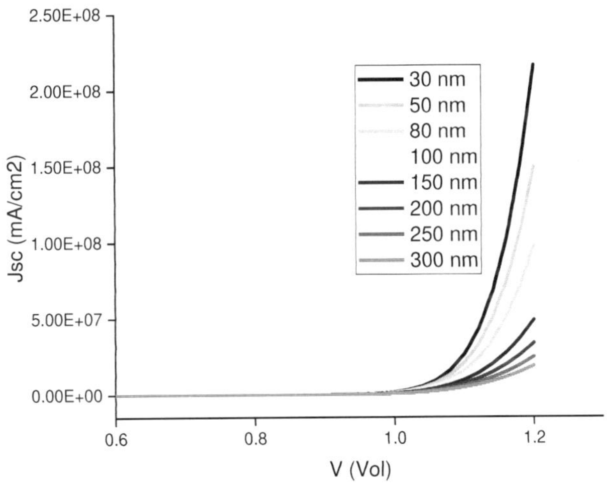

Figure 5. V_{oc} Vs. J_{sc} at different thicknesses of the $CH_3NH_3BiI_3$ perovskite layer.

This study demonstrates the impact of $CH_3NH_3BiI_3$ perovskite layer thickness on effectiveness, FF, J_{sc}, and V_{oc}. Thickness varied from 30 nm to 300 nm. The diffusion length must be greater than the cell thickness for the fluctuation between device parameters and absorber layer thickness, as depicted in figure 6. Due to defects, the minority charge carrier's diffusion length is typically relatively short, which causes the dominant charge separation to drift under the influence of the junction's electrostatic field, which cuts across the whole cell thickness. The FF and V_{OC} values significantly decrease as the thickness of its $CH_3NH_3BiI_3$ perovskite layer rises due to the increasing recombination rate of free charge carriers in the thicker photoactive active layer [14]. [14]. This is because as the thickness of the absorber layer increases, more photons are absorbed, but recombination rates also increase [15]. The FF decrease value for the $CH_3NH_3BiI_3$ perovskite layer thickness may be related to increased series resistance [24]. The photon absorption produces excitons that cannot cross the depletion layer. Due to the increased charge carrier recombination rate, the reverse saturation current becomes more significant, which causes the V_{oc} to decrease[14]. Furthermore, increasing exciton generation causes the PCE and J_{SC} to. Also, the increased thickness results in excessive resistance (with more defects and recombination rates), lowering essential values like the efficiency of the FPSC [11]. As thickness decreases, efficiency also does. As a result, we might conclude that a perovskite's considerably thicker layer may trap more photons, which results in many excitons. Therefore, we may conclude that a higher layer thickness could conversely increase the diffusion length and recombination rate, which would decrease PCE. The thickness of the absorption layer should be 100 nm, based on the simulation analysis.

4.1. Impact of Temperature on Device performance

The FPSC device's performance is influenced by temperature. The device's testing temperature is 300 K, whereas the operating temperature in the installed capacity is more than 300 K [25]. The simulated model's temperature ranged from 200 K to 500 K to analyse how temperature affects a solar cell's electrical performance. Figure 8 shows the temperature-related changes in the PCE. The efficiency of the simulated device is seen to decline with temperature, reaching 4.96% at 500 K.

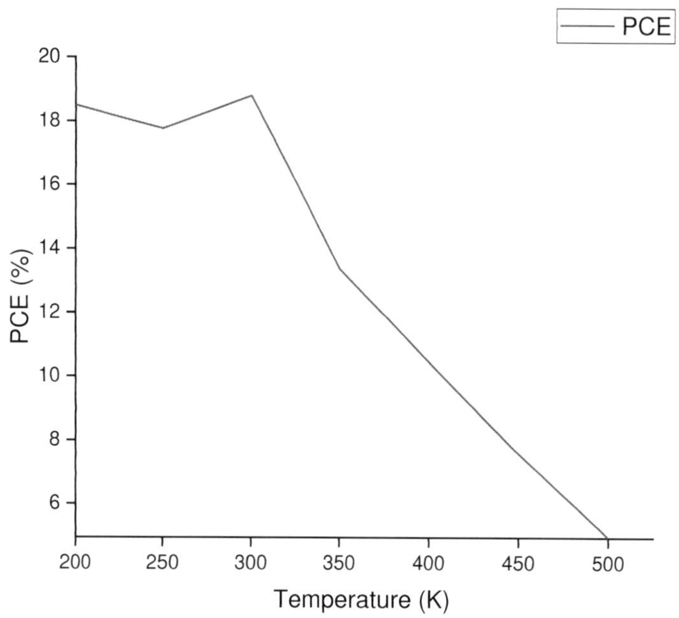

Figure 6. The graph of FPSC device efficiencies against temperatures.

Table 2. Various FF, V_{oc}, Jsc, and PCE of the simulated FPCS temperatures

Temperature (K)	V (Vol)	Jsc (mA/cm²)	FF (%)	PCE (%)
200	3.9123	24.1927	19.55	18.50
250	1.7636	24.1901	41.66	17.77
300	0.6332	35.5928	83.43	18.80
350	0.7525	24.1754	73.54	13.38
400	0.6284	24.1688	68.80	10.45
450	0.5033	24.2075	62.17	7.57
500	0.3788	24.2592	54.00	4.96

As a result of deformation and higher stress caused by the temperature increase, the layers may become more interconnected. The fill factor and efficiency will decrease as series resistance increases with a decrease in diffusion length [26]. The optimum temperature of the simulated model is determined to be 300K to obtain high efficiency, as illustrated in figure 8. All further optimization procedures were carried out at this temperature. With the temperature at 300 K, the design has a maximum efficiency of 18.80%, a fill factor of 83.43%, a Jsc of 35.59 mA/cm2, and a Voc of 0.332 V.

4.2. Impact of thickness on Defects.

By exploring the model equation is known as the Shockley-Read-Hall (SRH), the defect density (Nt) effect on the FPSC performance has been examined. The modelled simulated defect has a 0.58 eV mean energy between the band gap and a relatively neutral Gaussian distribution [24].

$$\text{SRH} = \frac{p.n - n_i^2}{\tau(n+ni) + \tau n(p+ni)} \qquad (6)$$

Equation 6 stands for "recombination rate," "p", and "n" stand for "pure concentrations of holes and electrons, respectively [14]. The model calculates diffusion length because SRH recombination is one of the most frequent types in PSC [11]. In this case, the defect density (Nt) value is 10^{15} cm^{-3} of ITO/PCBM/$MABiI_3$/Spiro-OMeTAD/Au, which is ideal for the device setup. The device's performance characteristics change as the active layer's (absorber layer) defect density (Nt) value rises, as displayed in figure 3 (b). The FPSC's defect density (Nt) must be reduced to an optimum amount to increase the efficiency of the perovskite device. During the simulation, removing the photoactive layer's defect density (Nt) is possible. Even these defects must be used to produce a good and effective FPSC device. Depending on the defect density, the solar cell's properties range from 10^{14} cm^{-3} to 10^{18} cm^{-3} (Nt)[16]. As was observed in table 2, any time the absorber layer has a high defect density value, the efficiency of the FPSC suffers dramatically due to an increased recombination rate and a decreased charge carrier lifetime, resulting in poor device performance [15].

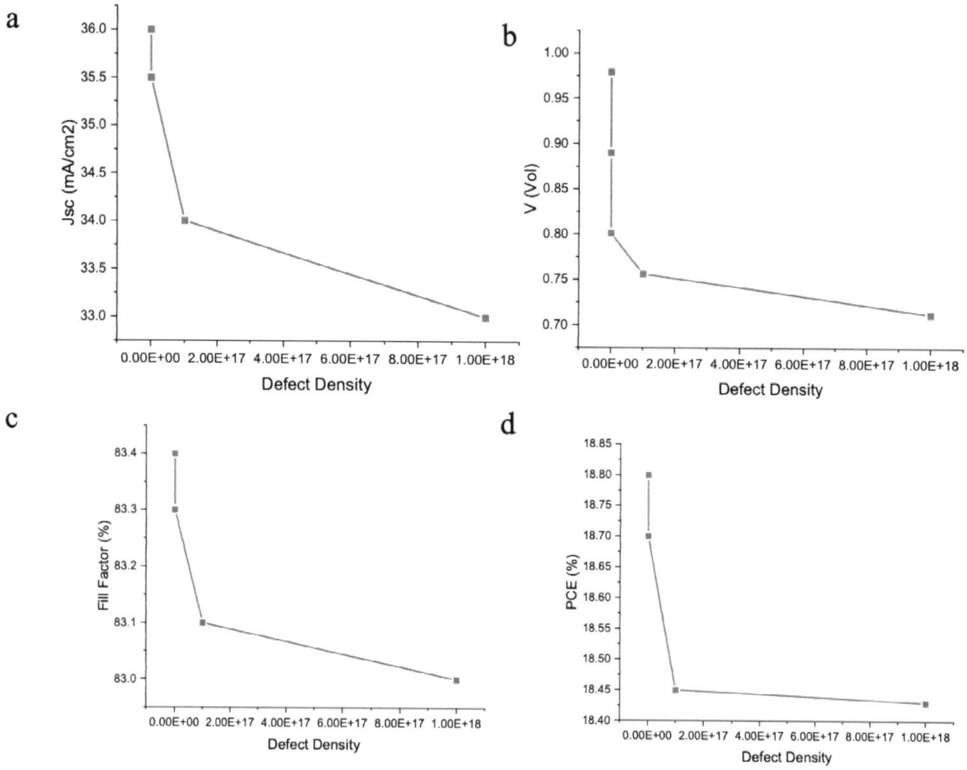

Figure 7. The impact of Nt of the $CH_3NH_3BiI_3$ absorber layer on the performance of (a) J_{sc}, (b) V_{oc} (c) FF, and (d) PCE.

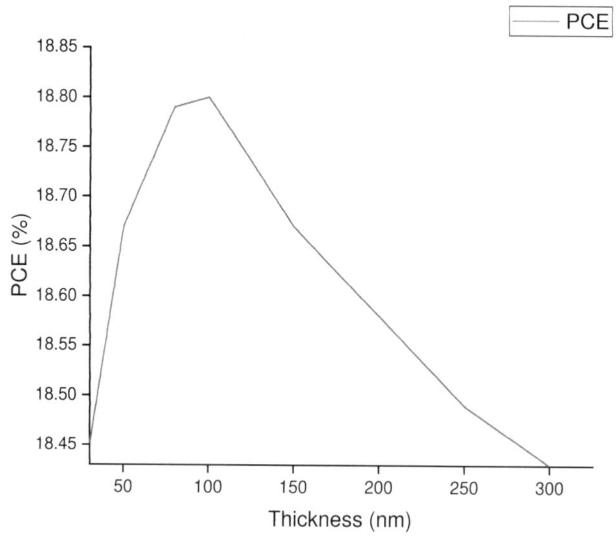

Figure 8. The graph of thickness vs. PCE of the simulated FPSC

Table 3. Various FF, V_{oc}, Jsc, and PCE of the simulated FPC

Thickness (nm)	V (Vol)	Jsc (mA/cm^2)	FF (%)	PCE (%)
30	0.6510	33.9674	83.42	18.45
50	0.6442	34.7638	83.33	18.67
80	0.6369	35.3763	83.41	18.79
100	0.6332	35.5928	83.43	18.80
150	0.6272	35.8222	83.12	18.67
200	0.6237	35.8513	83.08	18.58
250	0.6217	35.8138	83.06	18.49
300	0.6205	35.7624	83.04	18.43

Software was also used to numerically simulate and analyse the efficiency, defects, and the effect of thickness on FF in designing lead-free FPSC of the $CH_3NH_3BiI_3$ absorber layer. This solar cell has an architecture of ITO/PCBM/$CH_3NH_3BiI_3$/Spiro-OMeTAD/Au. Additionally, in the Bi-based FPSC, the thickness of the active layer significantly affects performance. The results show that many factors have an impact, such as the thickness of the absorber layer, FF, and defect density, as significant alteration was made to improve the efficiency of $MABiI_3$-based FPSC. The solar cell design with the highest efficiency, at a thickness of 100 nm, as shown in figure 7, had 18.80% PCE; an FF of 83.43%; and a J_{SC} of 35.59 mA/cm^2.

4.3. Impact of thickness on Fill Factor.

The PCE is affected by the FF, as shown in table 3, and is a significant parameter in perovskite solar cells, which is highly controlled by interfacial charge transfer processes; here, we have demonstrated its impact on the various thicknesses of the simulated FPSC, and it was observed as the thickness increases the FF factors decreases. Therefore, since FF is directly related to the device's PCE, a more moderate thickness is essential when fabricating FPSC devices.

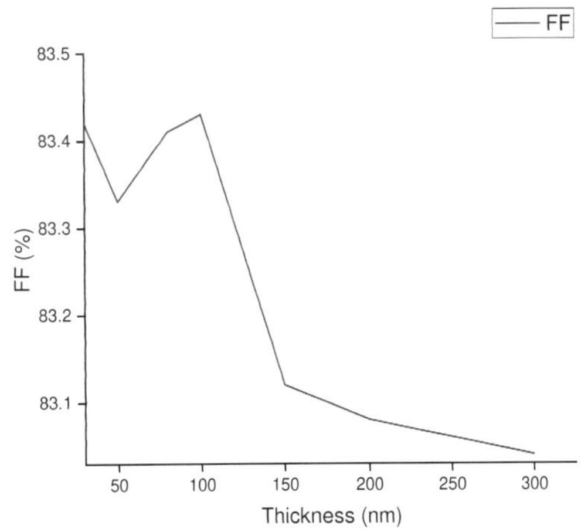

Figure 9. The graph of FF vs. thickness of the lead-free absorber layer

Table 4. Various FF and thicknesses of the simulated FPC

Thickness (nm)	FF (%)	PCE (%)
30	83.42	18.45
50	83.33	18.67
80	83.41	18.79
100	83.43	18.80
150	83.12	18.67
200	83.08	18.58
250	83.06	18.49
300	83.04	18.43

5. Conclusion

Design and simulation were performed for a lead-free FPSC. Using SCAPS-1D, we attempted to optimize the $CH_3NH_3BiI_3$ layer's thickness and connect the thickness to the FF. Using the architecture ITO/PCBM/$CH_3NH_3BiI_3$/Spiro-OMeTAD/Au for various thicknesses and relating the FF with the different thicknesses, we have simulated the FPSC model. With an absorber layer of 100 nm thick and a temperature of 300 K, the FPSC device has a maximum PCE of 18.80%. It was thus deduced that FF is indirectly proportional to the thickness of the FPSC. Additionally, we noted that for a 100 nm thick absorber layer, J_{SC} is 35.59 mA/cm^2, the V_{OC} is 0.6332 V, and the FF is 83.59%.

Acknowledgements

MOSTI-MIMOS Malaysia funded this research under Grant SRF-APP- RS-2021-003.

Reference:

[1] J. Di, J. Chang, and S. (Frank) Liu, "Recent progress of two-dimensional lead halide perovskite single crystals: Crystal growth, physical properties, and device applications," *EcoMat*, vol. 2, no. 3, Sep. 2020, DOI: 10.1002/eom2.12036.

[2] J. Di, J. Du, Z. Lin, S. Liu, J. Ouyang, and J. Chang, "Recent advances in resistive random access memory based on lead halide perovskite," *InfoMat*, vol. 3, no. 3. Blackwell Publishing Ltd, pp. 293–315, Mar. 01, 2021. DOI: 10.1002/inf2.12162.

[3] L. Li *et al.*, "Flexible all-perovskite tandem solar cells are approaching 25% efficiency with molecule-bridged hole-selective contact," *Nat Energy*, vol. 7, no. 8, pp. 708–717, Aug. 2022, DOI: 10.1038/s41560-022-01045-2.

[4] Y. Hu *et al.*, "Flexible Perovskite Solar Cells with High Power-Per-Weight: Progress, Application, and Perspectives," *ACS Energy Letters*, vol. 6, no. 8. American Chemical Society, pp. 2917–2943, Aug. 13, 2021. doi: 10.1021/acsenergylett.1c01193.

[5] X. Meng *et al.*, "Stretchable Perovskite Solar Cells with Recoverable Performance," *Angewandte Chemie - International Edition*, vol. 59, no. 38, pp. 16602–16608, Sep. 2020, doi: 10.1002/anie.202003813.

[6] T. A. Berhe *et al.*, "Organometal halide perovskite solar cells: Degradation and stability," *Energy and Environmental Science*, vol. 9, no. 2. Royal Society of Chemistry, pp. 323–356, Feb. 01, 2016. DOI: 10.1039/c5ee02733k.

[7] N. K. Noel *et al.*, "Lead-free organic-inorganic tin halide perovskites for photovoltaic applications," *Energy Environ Sci*, vol. 7, no. 9, pp. 3061–3068, 2014, DOI: 10.1039/c4ee01076k.

[8] L. Fu, Y. Nie, B. Li, N. Li, B. Cao, and L. Yin, "Bismuth Telluride Interlayer for All-Inorganic Perovskite Solar Cells with Enhanced Efficiency and Stability," *Solar RRL*, vol. 3, no. 12, Dec. 2019, DOI: 10.1002/solr.201900233.

[9] N. F. Coutinho *et al.*, "The Thermomechanical Properties of Thermally Evaporated Bismuth Triiodide Thin Films," *Sci Rep*, vol. 9, no. 1, Dec. 2019, DOI: 10.1038/s41598-019-48194-1.

[10] H. X. Zhu and J. M. Liu, "Electronic structure of organometal halide perovskite CH3NH3BiI3 and optical absorption extending to infrared region," *Sci Rep*, vol. 6, Nov. 2016, DOI: 10.1038/srep37425.

[11] K. Tan, P. Lin, G. Wang, Y. Liu, Z. Xu, and Y. Lin, "Controllable design of solid-state perovskite solar cells by SCAPS device simulation," *Solid State Electron*, vol. 126, pp. 75–80, Dec. 2016, DOI: 10.1016/j.sse.2016.09.012.

[12] L. Lin, L. Jiang, Y. Qiu, and Y. Yu, "Modeling and analysis of HTM-free perovskite solar cells based on ZnO electron transport layer," *Superlattices Microstruct*, vol. 104, pp. 167–177, Apr. 2017, DOI: 10.1016/j.spmi.2017.02.028.

[13] N. Singh, A. Chaudhary, S. Saxena, M. Saxena, and N. Rastogi, "Electrical Simulation of Organic Solar Cell at Different Charge Carrier Mobility," *IOSR Journal of Applied Physics*, vol. 09, no. 02, pp. 01–04, Apr. 2017, doi: 10.9790/4861-0902020104.

[14] U. Mandadapu, "Simulation and Analysis of Lead based Perovskite Solar Cell using SCAPS-1D," *Indian J Sci Technol*, vol. 10, no. 1, pp. 1–8, Jan. 2017, doi: 10.17485/ijst/2017/v11i10/110721.

[15] K. Deepthi Jayan and V. Sebastian, "Comprehensive device modelling and performance analysis of MASnI3 based perovskite solar cells with diverse ETM, HTM, and back metal contacts," *Solar Energy*, vol. 217, pp. 40–48, Mar. 2021, DOI: 10.1016/j.solener.2021.01.058.

[16] A. Kumar and S. Singh, "Numerical modeling of lead-free perovskite solar cell using inorganic charge transport materials," in *Materials Today: Proceedings*, 2019, vol. 26, pp. 2574–2581. doi: 10.1016/j.matpr.2020.02.545.

[17] M. Burgelman, K. Decock, A. Niemegeers, J. Verschraegen, and S. Degrave, "SCAPS manual."

[18] F. Azri, A. Meftah, N. Sengouga, and A. Meftah, "Electron and hole transport layers optimization by numerical simulation of a perovskite solar cell," *Solar Energy*, vol. 181, pp. 372–378, Mar. 2019, DOI: 10.1016/j.solener.2019.02.017.

[19] A. Slam and A. B. Belkaid, "Numerical Study of Based Perovskite Solar Cells by SCAPS-1D Solar cells View project Solar cells View project." [Online]. Available: https://www.researchgate.net/publication/337797426

[20] G. Haidari, "Comparative 1D optoelectrical simulation of the perovskite solar cell," *AIP Adv*, vol. 9, no. 8, Aug. 2019, DOI: 10.1063/1.5110495.

[21] Y. Kuang *et al.*, "Low-Temperature Plasma-Assisted Atomic-Layer-Deposited SnO2 as an Electron Transport Layer in Planar Perovskite Solar Cells," *ACS Appl Mater Interfaces*, vol. 10, no. 36, pp. 30367–30378, Sep. 2018, doi: 10.1021/acsami.8b09515.

[22] L. Lin, L. Jiang, Y. Qiu, and Y. Yu, "Modeling and analysis of HTM-free perovskite solar cells based on ZnO electron transport layer," *Superlattices Microstruct*, vol. 104, pp. 167–177, Apr. 2017, DOI: 10.1016/j.spmi.2017.02.028.

[23] K. Tan, P. Lin, G. Wang, Y. Liu, Z. Xu, and Y. Lin, "Controllable design of solid-state perovskite solar cells by SCAPS device simulation," *Solid State Electron*, vol. 126, pp. 75–80, Dec. 2016, DOI: 10.1016/j.sse.2016.09.012.

[24] A. Bag, R. Radhakrishnan, R. Nekovei, and R. Jeyakumar, "Effect of absorber layer, hole transport layer thicknesses, and its doping density on the performance of perovskite solar cells by device simulation," *Solar Energy*, vol. 196, pp. 177–182, Jan. 2020, DOI: 10.1016/j.solener.2019.12.014.

[25] S. R. Raga and F. Fabregat-Santiago, "Temperature effects in dye-sensitized solar cells," *Physical Chemistry Chemical Physics*, vol. 15, no. 7, pp. 2328–2336, Feb. 2013, DOI: 10.1039/c2cp43220j.

[26] A. Slam and A. B. Belkaid, "Numerical Study of Based Perovskite Solar Cells by SCAPS-1D Solar cells View project Solar cells View project." [Online]. Available: https://www.researchgate.net/publication/337797426

Device Optimization of a Pb-Free All Perovskite Tandem Solar Cell with 29.59% Power Conversion Efficiency

Mohammad Aminul Islam[1*], Mhd Deb Abou Hashish[1], S F W Muhammad Hatta[1], Norhayati Binti Soin[1], Sobayel Khan[2], Nowshad Amin[3]

[1]Department of Electrical Engineering, Faculty of Engineering, Universiti Malaya, Jalan Universiti, 50603 Kuala Lumpur, MALAYSIA.
[2]Solar Energy Research Institute (SERI), Universiti Kebangsaan Malaysia, 43600 Bangi, Selangor, MALAYSIA.
[3]Institute of Sustainable Energy, Universiti Tenaga Nasional, Jalan IKRAM-UNITEN, 43000 Kajang, Selangor, MALAYSIA
*Corresponding author: aminul.islam@um.edu.my

Abstract. This study has focused on simulating a highly efficient lead-free all-perovskite tandem solar cell using the SCAPS 1D device simulation tool. In the tandem structure, the top cell used Cs2AgBi0.75Sb0.25Br6 (Eg = 1.80 eV), and the bottom cell used FAMASnGeI3 (Eg = 1.40 eV) as the absorber material. Also, ZnO and NiOx were used as the electron transport layer (ETL) and the hole transport layer (HTL), respectively. Primarily the individual top cell and bottom cell have been optimized. The highest efficiency of the top cell was found to be 17.13% with performance parameters V_{OC} = 1.23V, J_{SC} = 15.57 mA/cm^2 and FF = 89.39%, whereas the optimized efficiency of the bottom cell was found to be 17.58% with performance parameters V_{OC} = 0.85V, J_{SC}= 27.38 mA/cm^2 and FF = 75.34%. The thickness of the absorber of the top cell and bottom cell shows a significant impact on the device performance and the optimum thickness for the $Cs_2AgBi_{0.75}Sb_{0.25}Br_6$ absorber layer was found to be 400 nm, whereas for the $FAMASnGeI_3$ layer was found to be 200 nm. Careful optimization of the tandem device has resulted in an improvement of the performance and obtained an efficiency of 29.59 %, with J_{SC} of 17.50 mA/cm^2, V_{OC} of 2.09 V, and FF of 80.87%. The final simulated device sums up tremendous potential for the fabrication of a highly efficient PSC device using lead-free perovskite materials towards excelling commercialization.

1. Introduction

Over the past few decades, solar photovoltaic (PV) technology has advanced quickly due to technological advancements, the development of novel materials, and an increase in commercial manufacturing [1]. Also, energy demand over the world also increasing simultaneously. To mitigate the energy demand and decrease reliance on fossil fuels, the efficiency of the PV solar cell should exceed 30%. However, it is not possible to achieve such efficiency using single junction solar cells as was confirmed by the Shockley–Queisser (SQ) limit [2]. The performance is mainly limited by two factors; first, the optical loss; where the photon energies lower than the bandgap cannot be absorbed, and second, thermalization loss. The thermalization losses occur when photon energy exceeds that needed to form free electron-hole pairs, causing carriers to thermalize instead of contributing to the device's usable output. These losses can be reduced by connecting two or more solar cells of the antithetic PV materials of the appropriate bandgap. In that case, the whole incident spectrum will be separately absorbed by the stacks. To effectively use the full sun spectrum from lower to higher wavelength photons, the material with wide and narrow bandgap is utilized for the top and bottom subcell in this approach [3]. Several researchers conducted theoretical and experimental research on multijunction concentrator cells, also known as cascade or tandem cells. [4, 5]. There is numerous types of tandem solar cells have already been researched and commercialized. However, their performance and the commercial price are not yet at the expected level. Alternative to other types of solar cells, perovskite solar cells (PSCs) are showing a prominent interest in solar cell technologies because of their encouraging performance and easiest fabrication technique [6]. One of the favorite nominees of the perovskite family that has been extensively investigated by multiple studies is CH3NH3PbI3. Nonetheless, the high toxicity of lead in this type of material restricts its usage in real-life applications in the industry. And forces on searching for alternative lead-free perovskite materials. Also, the efficiency of solar cell devices has been of great interest in hoping for reaching a very high efficiency. Therefore, this project has focused on simulating a highly efficient Pb-free all-perovskite tandem solar cell (PTSC) using the SCAPS 1D device simulation tool. Table 1 shows a few of the recent works alongside the reported highest performance.

Table 1. Different studies of different perovskite materials used in tandem solar cell

Author and date	Absorber layer materials (top and bottom cell)	Band gap (eV) (top and bottom cell)	Reported efficiency
Amri et al.,2021 [7]	Cs2AgBi0.75Sb0.25Br6 and n-Si	1.8 – 1.12	24.4%
Madan et al.,2020 [8]	Cs2AgBi0.75Sb0.25Br6 And FACsPb0.5Sn0.5I3	1.8 – 1.2	17.35%
Singh et al.,2020 [9]	CH3NH3GeI3 and FA0.75MA0.25Sn0.25Ge0.5I3	1.9 – 1.4	26.72%
Pandey et al.,2021 [10]	Cs2AgBi0.75Sb0.25Br6 And MAPb0.5Sn0.5I3	1.82 – 1.22	22%

This study includes the simulation of a novel structure Pb-free PTSC based upon double perovskite Cs2AgBi0.75Sb0.25Br6 (Eg = 1.80 eV) and FAMASnGeI3 (Eg = 1.40 eV). The Cs2AgBi0.75Sb0.25Br6 have proven some preferable advantages as double halide perovskites, such as superior stability in open-air conditions and suitable bandgaps as top sub-cell in tandem solar cells [11]. The effect of both absorber layers' thickness on the performance of the standalone solar cell has been investigated. We also tried to find out the optimal ETL combination with the proposed PTSC. It should be noted that the ETL layer is indispensable for the proper facilitation of electrons from the perovskite absorber layer to the electrodes. One-dimensional SCAPS (Solar Cell Capacitance Simulator version 3.3.07) tool has been used to simulate and investigate the device. SCAPS is widely utilized to numerically analyzed solar cells due to its innate capability for simulating defect configurations (bulk and interface) as well as energy distributions including uniform, single, or Gaussian defects. Moreover, it has allowed changing the work function of the contact materials and to use of optical filters for reflection and transmission functions. Several publications also indicate a successful application of the PSCs employing SCAPS 1D with outstanding PV performance [13-15]. Therefore, numerical modeling of PSCs using SCAPS-1D is a suitable method for its' progressing advancement in the field of PV.

2. Device Structure and Parameters

At first, the device performance of individual and standalone PSCs based on Cs2AgBi0.75Sb0.25Br6 and FAMASnGeI3 absorber layers has been studied to provide insight into the PV performance of the lead-free PTSC. The device structures for standalone PSCs and PTSC have been shown in Figure 1(a), (b) and (c).

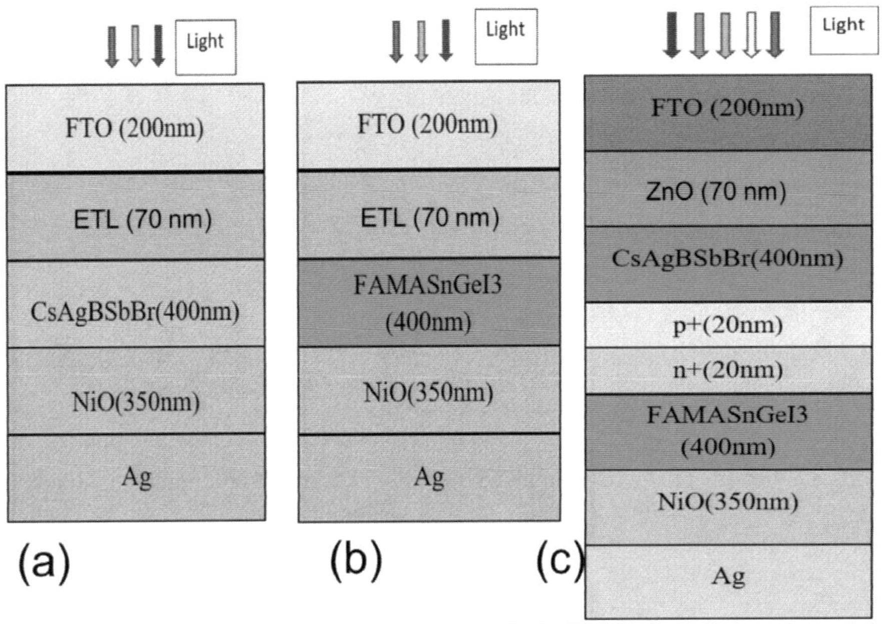

Figure 1. Simulated structure of (a) Cs2AgBi0.75Sb0.25Br6, (b) FAMASnGeI3-based standalone PSC, and (c) proposed PTSC.

Table 2. The initial layer parameters adopted for this study (symbols have their usual meanings) [8, 16-20]

Parameters	NiO	FAMASnGeI3	$Cs_2AgBi_{0.75}Sb_{0.25}Br_6$	ZnO	FTO
Eg (eV)	3.8	1.40	1.80	3.3	3.50
Xe (eV)	1.46	3.67	3.98	4.0	4.00
w (nm)	350	400	400	70	200
μn (cm^2/Vs)	2.8	2	1.62×10^5	100	20
μp (cm^2/Vs)	2.8	2	1.010×10^5	25	10
εr	11.7	8.20	10.0	9.0	9.00
Nc (cm^{-3})	2.5×10^{20}	2.2×10^{18}	1×10^{16}	3.7×10^{18}	2.2×10^{18}
Nv ((cm^{-3})	2.5×10^{20}	1.8×10^{19}	1×10^{16}	1.8×10^{19}	1.8×10^{19}
$v_{e,th}$ (cm/s)			1×10^7		
$v_{h,th}$ (cm/s)			1×10^7		
Na (cm^{-3})	3×10^{18}	0	1×10^9	0	0
Nd (cm^{-3})	0	1×10^{13}	1×10^9	5×10^{17}	1×10^{18}

Table 2 shows the parameters of the different active layers which we have adopted from the different published works [8, 16-20] for the simulation of single-layer stand-alone PSCs and PTSC. We primarily calibrate the top and bottom solar cells and analyze their photovoltaic (PV) outputs. Then, we investigate how the PV parameters evolved with the thickness of higher and lower bandgap perovskite materials. Finally, the absorber layer (top and bottom sub-cell) has been tuned in such a way that the will produce almost similar J_{SC}. These optimized top and bottom cells are then stacked together in a 2T tandem connection as shown in Figure 1(c) and investigate the tandem performance. From Figure 1, it can be seen that the light spectrum will pass through the electron transport layer (ETL) and this also makes a bridge between the front contact and the absorber layer, thus it is essential to find out the best-suited material as an ETL.

3. Results and discussion

3.1. Calibration of top cell and bottom cell

Stand-alone PSCs with the wide band–gap (Cs2AgBi0.75Sb0.25Br6) and narrow band–gap (FAMASnGeI3) single absorber layers were initially modelled as a top and bottom cell. The J-V (current density-voltage) curve as well as the EQE (external quantum efficiency) found from the simulation are shown in Figure 2. (a-d). As shown in Figure 2(a), the Voc is 1.23V, Jsc is 15.57 mA/cm2, and the efficiency is 17.13%, whereas these values for the bottom sub-cell are Voc = 0.85V, Jsc = 27.38 mA/cm2, and efficiency = 17.58%, respectively. Moreover, at a wavelength of 400 nm, both cells exhibited EQE of greater than 80%, however, the EQE of the top cell fell below 60% for wavelengths greater than 650 nm. This is because of the high bandgap of the perovskite solar cell used as the top cell (1.8 eV). The computed EQE for a 1.8 eV top cell is extremely similar to the experimental device reported elsewhere [21]. Figure 2(b) and 2(d) indicate that the bottom subcell has a cut-off wavelength for EQE that is about 50% higher, together with a greater spectrum width, leading to improved spectrum utilization and, consequently, higher Jsc has been achieved. In addition, the insets of Figures 2(a) and (c) offer a summary of the PV parameters of the simulated devices, including Voc, Jsc, FF, and power conversion efficiency (PCE).

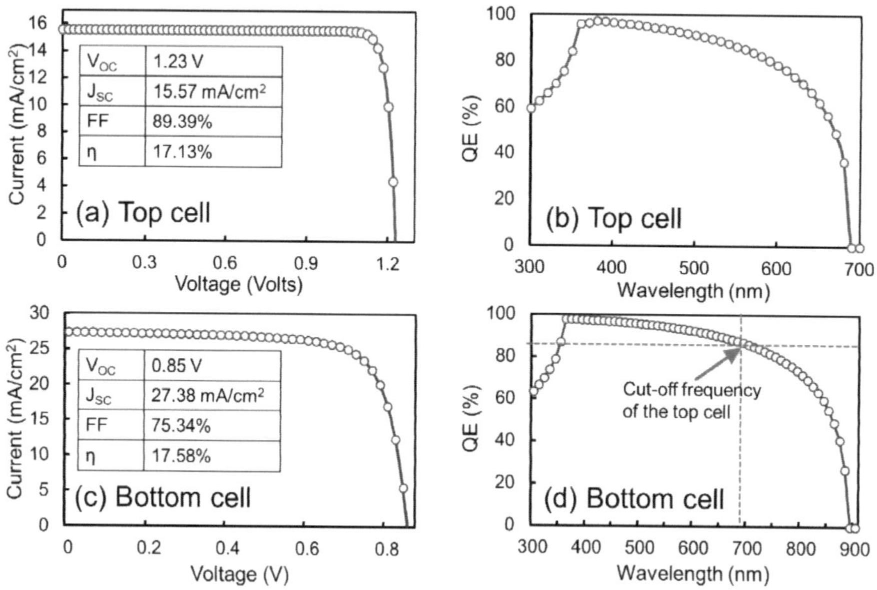

Figure 2. Simulated J-V curve ((a) and (c)) and QE ((b) and (d)) of the top and bottom stand-alone sub-cell

The energy band diagram (EBD) of the top and bottom cells has also been acquired at zero bias conditions under illumination to comprehend the carrier dynamics within and the overall performance of the device. Figure 3(a) and (b) illustrate the EBD of the top and bottom subcells at zero bias conditions. With the aid of EBD, the results clearly demonstrate that the top cell's higher V_{OC} value is valid, as shown in Fig. 4(a) and (b). Note that the Quasi-Fermi Level (QFL) offset to determine the Voc of the PV devices, and the offset is found to be higher for the top cell than the bottom cell which confirmed the higher Voc. Also, the measured Voc (1.23 V for the top cell and 0.85 V for the bottom cell) was used in both devices to figure out how the QFL offset was changing over time. This proved that QFL flattened towards the perovskite/ETL interface (not shown here) and did not affect Voc.

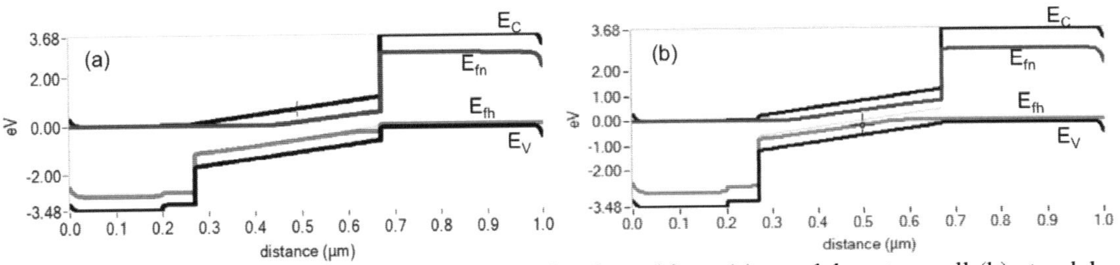

Figure 3. Energy band diagram with zero applied voltage bias, (a) standalone top cell (b) standalone bottom cell

3.2 Absorber layer optimization:

The top and bottom cells were assessed in standalone settings before the performance of the PTSC were examined, and the effect of the absorber layer thickness was investigated. Figures 4(a) and (b) show the PV parameters at various thicknesses of the absorber layer for both stand-alone solar cells. It can be observed that increasing the active layer thickness has an evidential impact on the device's performance. Also, Figure 4(a) shows that the device performance is increased until the thickness of 400 nm of the top cell's absorber layer. Quantitatively, the average increase of J_{SC} is 13.37% and an efficiency of 12.85% with a marginal improvement of V_{OC} and FF was observed for the top cell. The betterment is attributed to improved optical absorption in the device's active region, in which the electron-hole pairs are promoted. Subsequently, the J_{SC} and other performance parameters are improved. For the bottom cell, device performance is also increased until 200 nm of the $FAMASnGeI_3$ absorber layer.

Moreover, the average increase of J_{SC} is 5.27%, FF is 6.83% and efficiency of 12.85% with no improvement of V_{OC} was observed for the bottom cell. Both cells are showing decreasing trends of the performance parameters after crossing a certain thickness. The decline in performance can be attributed to an increase in resistance as absorber thickness increases [22]. It should be noted that the overall series resistance of the device can be increased with the increase of active layer thickness. Besides, because the charges must travel a longer distance for diffusion in a thicker absorber layer, there is a possibility to increase carrier recombination. As a result, efficiency declines after a particular thickness value. These results are quite consistent with the experimental findings of Correa-Baena et al. [23]. V_{OC} and other parameter declination are caused by accelerated recombination with increasing thickness. Hence, 400 nm and 300 nm can be chosen as the ideal thickness for the $Cs_2AgBi_{0.75}Sb_{0.25}Br_6$ and $FAMASnGeI_3$-based perovskite layer. Similar results for different structures of $FAMASnGeI_3$-based solar cells have also been reported elsewhere [8].

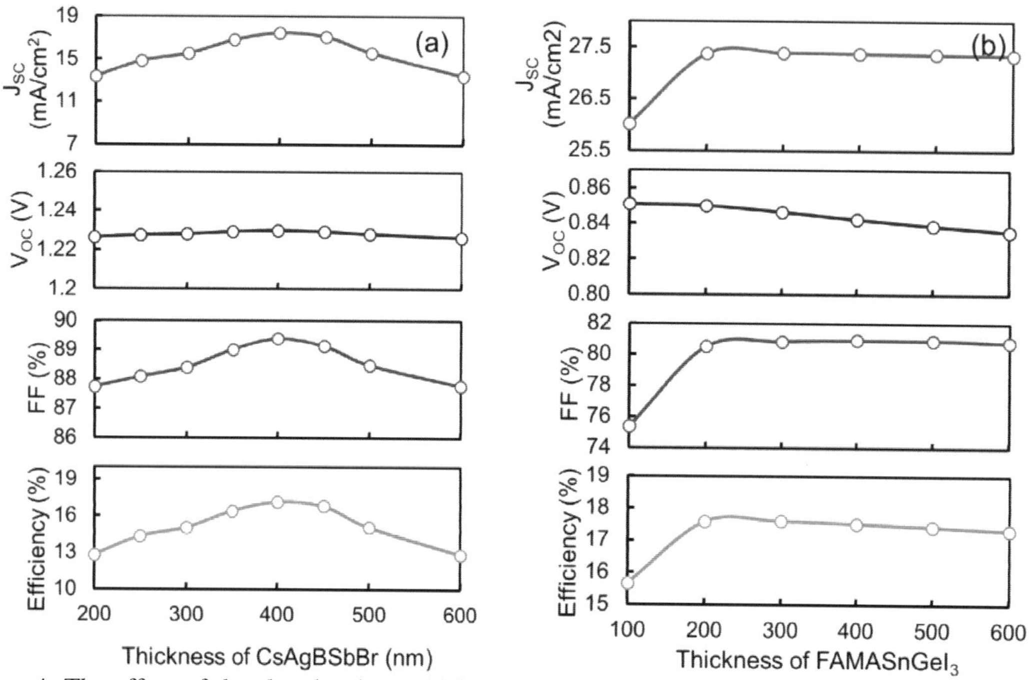

Figure 4: The effect of the absorber layer thickness on the performance of the solar cell, (a) top cell and (b) bottom cell

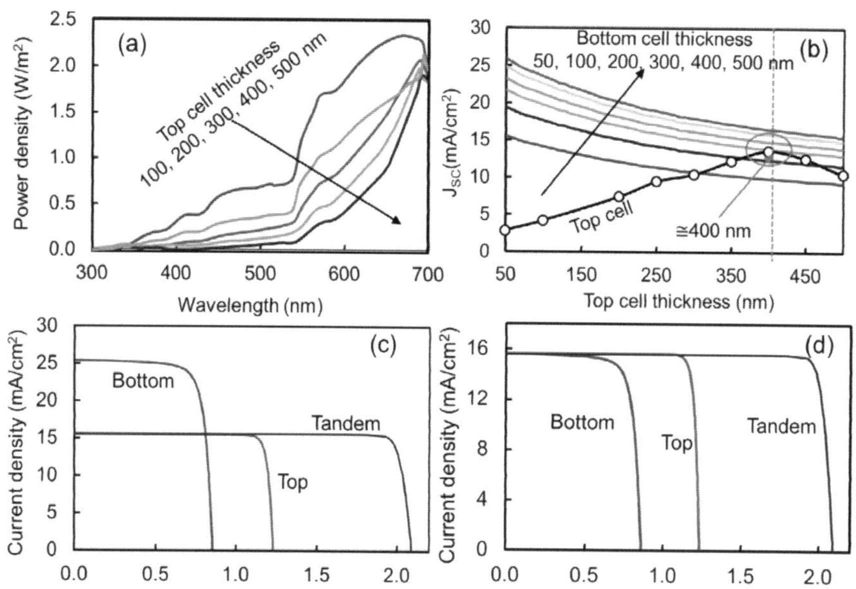

Figure 5(a) Transmitted spectrum for various thicknesses of the absorber of the top sub-cell, (b) Jsc for top and bottom cells at various top and bottom perovskite material thicknesses (50–500 nm)(the optimum paring is at around 390 nm of top cell perovskite material indicated by arrow sign), (c) J-V curve of the tandem cell and standalone top and bottom sub-cell, and (d) J-V curves of the tandem cell as well as top and bottom cell after filtered spectrum feeding.

After the optimization of the individual solar cells, they are connected in series for tandem application as shown in Figure 1(c). The I-V curve found in the simulation is depicted in Figure 5(c). Figure 5(c) shows that the tandem solar cell's current density is nearly equal to the current density of the top cell, whereas open circuit voltage is the sum of the two sub-cells. It should be noted that the electrical behavior of two-terminal tandem solar cells is similar to that of two diodes linked in series, therefore an equal amount of current must always flow through each cell. Consequently, the cell with lower J_{SC} values (here top cell) serves as a limiting cell for the tandem device's overall J_{SC}. Besides, the total voltage drop across the tandem device is the sum of the individual voltage drops of each cell[24]. Accordingly, the performance of the tandem device is greatly influenced by the thickness of the active layer in the top and bottom cells. Consequently, top and bottom cell mismatches result in the subpar performance of the tandem device. The thickness of the top and bottom cells must be adjusted for the tandem device's best performance in order to achieve equal Jsc and permit identical tunnel recombination junctions [25-28]. If parasitic absorption increases above the optimum level in the top cell and optical coupling failed in the bottom cell, the tandem device's overall Jsc will be reduced. Another option is to reduce the top cell thickness below the optimal value, which has the same effect as reducing Jsc by reducing the top's absorption. In this research, the filtered spectrum approach is used to obtain current matching circumstances that result in a uniform Jsc value between the two cells.

Table 3. Solar cell output parameters of the tandem cell as well as top and bottom cell after feeding with the filtered spectrum.

Cell	Jsc (mA/cm^2)	Voc (V)	FF (%)	PCE (%)
Top cell	15.58	1.21	88.98	17.21
Bottom cell	15.57	0.86	77.67	17.43
Tandem cell	15.58	2.09	80.87	29.59

Here, the top cell is lighted with the AM1.5 spectrum, and the transmitted spectrum, S(λ), by the top cell is computed using the absorption coefficient and thickness of all the layers utilizing the following equation.

$$S(\lambda) = S_0(\lambda).\exp\,[-\alpha(\lambda).d] \qquad (1)$$

Here, "$S_0(\lambda)$" represents the incident spectrum at AM1.5, "α" represents the absorption coefficient and "d" represents the perovskite material thickness of the top cell. For simplicity, the interfacial reflection losses as well as absorption by other layers, such as NiO and FTO have been ignored. Figure 5(a) shows the estimated transmitted spectrum after being filtered by top cells with different thicknesses. It is evident that the power of the filtered transmitted spectrum is decreased, notably below the wavelengths of 700 nm, as the absorber layer in the top cell becomes thicker. This is due to increased absorption in top cells with thick absorber layers.

To investigate the optical coupling of photons in the two sub-cells, the estimated filtered spectra have been utilized. The bottom cell is received higher photocurrent when the thickness of the top subcell's perovskite layer thickness is less. It has been found that the transmitted photocurrent decreases as the top sub-cell thickness approach 400 nm. As previously stated, the overall Jsc of the tandem cell decreases as the thickness of the top-subcell decreases below a certain threshold value. Similarly, if the top cell's thickness rises above a certain level, the absorption in the bottom cell will decrease and overall J_{SC} reduce again. Following this fact, we established a current matching condition to acquire the same J_{SC} value in both sub-cells. We used the filtered spectra as shown in Figure 5(a) for illuminating the bottom cell for evaluating the PV parameters. The thickness of the bottom subcell varied from 50 to 500 nm. The J_{SC} values of the bottom sub-cell obtained in this process as shown in Figure 5(b) were utilized to investigate the pairing status of the tandem device. The best optical coupling or current matching condition which confirms the higher equal J_{SC} for both sub-cells is found to be a top and bottom sub-cell thickness of 390 nm and 200 nm, respectively yielding J_{SC} values equal to 15.58 mA/cm^2 and 15.57 mA/cm^2. The PV parameters of both sub-cells under the current matching condition are shown in Table 2. Finally, both sub-cell is connected in series (2T) as shown in Figure 1(c) and the simulated J-V curves for the stand-alone condition and J_{SC} matched condition are shown in Figure 5(c) and 5(d). From the simulation, the highest efficiency for the tandem is found to be, V_{OC} = 2.09 V, J_{SC} = 17.50 mA /cm^2, FF = 80.87%, and efficiency = 29.59%.

3.3 Effect of Temperature

Solar cells are typically employed outdoors where they are exposed to sunlight, which can raise the cell temperature by as much as 60°C and sometimes even more depending on the season and location [29]. For understanding of long-term stability related with the temperature, readers are referred to the work reported elsewhere [30, 31]. As the temperature impacts are crucial, we simulated the working temperature's impact on the proposed tandem solar cell.

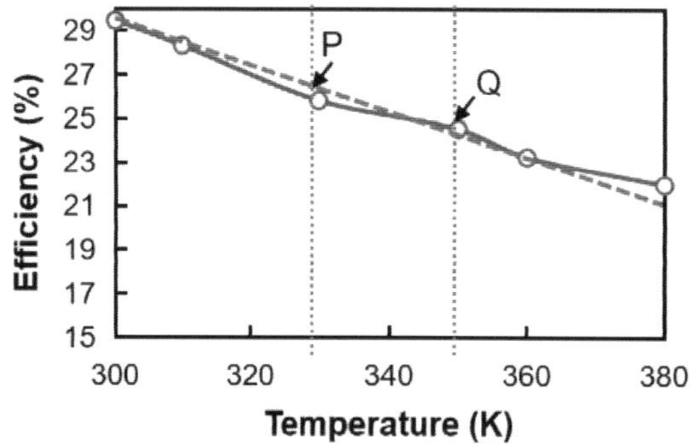

Figure 6 Effect of the working temperature on the efficiency of the tandem solar cell (P and Q, two points selected to calculate the temperature gradient)

The effect of the working temperature on the efficiency of the tandem solar cell can be seen in Figure 6. The rising temperature primarily affects the V_{OC} and its decay with the increase of temperature which leads to the decrease of efficiency of the solar cell. Particularly, due to its direct relationship with reverse saturation current density (J_o), which is also connected to intrinsic carrier concentration (n_i), V_{OC} is decreased. Additionally, a high temperature may generate more excited electrons, which would increase carrier recombination and lower efficiency. The temperature coefficient has also been calculated from Figure 6 and it has been found that the coefficient is -0.098%/°C for the proposed all perovskite Pb-free perovskite solar cells.

4. Conclusion

The state-of-the-art conversion efficiencies of the individually operated $Cs_2AgBi_{0.75}Sb_{0.25}Br_6$ and $FAMASnGeI_3$ perovskite absorber layer with bandgap 1.4 eV and 1.8 eV, respectively, are simulated and calibrated. Substantial efforts were also given to design an all-perovskite Pb-free tandem device with the help of these perovskite absorber materials. To achieve a realistic tandem operation, the absorber layer thickness in both the top and bottom subcells has been altered to attain the current matching conditions. For this, the top cell has been illuminated by the standard AM1.5 spectrum, while the bottom cell was simultaneously investigated using the transmitted spectrum coming through the top cell. In the current matching condition, the optimum thickness for the top and bottom cell absorber layer was found to be 390 nm and 200 nm, respectively. The efficiency given by individual cells is 17.21% (top) and 17.43% (bottom) under the current matching condition. For achieving a tandem device, the top and bottom cells are connected in a series (2T) connection. The proposed tandem device shows an attractive power conversion efficiency of 29.59% including PV parameters V_{OC} = 2.09 V, J_{SC} = 17.50 mA /cm^2, and FF = 80.87%, respectively. As the perovskite-perovskite tandem solar cell is in its infancy at this time, this work lays the groundwork for the future development of non-toxic solar cells with high efficiency by providing a comprehensive simulation of a Pb-free all-perovskite tandem architecture. Future research into the effect of tunnel recombination junctions and structural and parametric improvements to the reported device offer scopes for even higher conversion efficiencies.

Acknowledgment: The authors wish to express their deepest appreciation to the Universiti Malaya for its support through Grand Project No: RMF0423-2021. This work was also partially supported by the Malaysian ministry of higher education through the FRGS grant FRGS/1/2020/TK0/UM/02/33. The authors also acknowledge the Faculty of Engineering, Universiti Malaya (@UM) for other support.

References

[1] S. T. Mohammad, H. H. Al-Kayiem, M. A. Aurybi, and A. K. Khlief, "Measurement of global and direct normal solar energy radiation in Seri Iskandar and comparison with other cities of Mala[1] Reyes-Belmonte, M. A. (2021). Quo
vadis solar energy research?. Applied Sciences, 11(7), 3015.

[2] Shockley, W., & Queisser, H. J. (1961). Detailed balance limit of efficiency of p-n junction solar cells. Journal of applied physics, 32(3), 510-519.

[3] Jackson, E.D., 1955. Areas for improvement of the semiconductor solar energy converter. Trans. Intern. Conf. on the Use of Solar Energy-The. Scient. Basis 5, 122.

[4] Loferski, J., 1982. Theoretical and experimental studies of tandem or cascade solar cells-A review. In: 16th Photovoltaic Specialists Conference. pp. 648–654.

[5] Hutchby, J., 1985. A review of multijunction concentrator cells. In: Proc. 18th IEEE Photovoltaic Specialists Conf. pp. 20–27.

[6] Jiang, F., Liu, T., Luo, B., Tong, J., Qin, F., Xiong, S., Li, Z., Zhou, Y., 2016. A twoterminal perovskite/perovskite tandem solar cell. J. Mater. Chem. A 4, 1208–1213.

[7] Amri, K., Belghouthi, R., Aillerie, M., & Gharbi, R. (2021). Device optimization of a lead- free perovskite/silicon tandem solar cell with 24.4% power conversion efficiency. Energies, 14(12), 3383.

[8] Madan, J., Shivani, Pandey, R., & Sharma, R. (2020). Device simulation of 17.3% efficient lead-free all-perovskite tandem solar cell. Solar Energy, 197, 212–221.

[9] Singh, N., Agarwal, A., & Agarwal, M. (2020). Numerical simulation of highly efficient lead-free all-perovskite tandem solar cell. Solar Energy, 208, 399–410.

[10] Pandey, R., Sharma, S., Madan, J., & Sharma, R. (2021). Numerical simulations of 22% efficient all-perovskite tandem solar cell utilizing lead-free and low lead content halide perovskites. Journal of Micromechanics and Microengineering, 32(1), 014004

[11] Kumar, A., & Singh, S. (2021). Computational simulation of metal doped lead-free double perovskite (Cs2AgBi0. 75Sb0. 25Br6) solar cell using solar cell capacitance simulator. Materials Today: Proceedings, 44, 2215-2222.

[12] Duan, Q., Ji, J., Hong, X., Fu, Y., Wang, C., Zhou, K., Liu, X., Yang, H., Wang, Z.Y., 2020. Design of hole-transport-material free CH3NH3PbI3/CsSnI3 all-perovskite heterojunction efficient solar cells by device simulation. Sol. Energy 201, 555–560.

[13] Zhao, P., Liu, Z., Lin, Z., Chen, D., Su, J., Zhang, C., Zhang, J., Chang, J., Hao, Y., 2018b. Device simulation of inverted CH3NH3PbI3□ xClx perovskite solar cells based on PCBM electron transport layer and NiO hole transport layer. Sol. Energy 169, 11–18.

[14] Chakraborty, K., Choudhury, M. G., & Paul, S. (2019). Numerical study of Cs2TiX6 (X= Br–, I–, F– and Cl–) based perovskite solar cell using SCAPS-1D device simulation. Solar Energy, 194, 886-892.

[15] Raoui, Y., Ez-Zahraouy, H., Tahiri, N., El Bounagui, O., Ahmad, S., Kazim, S., 2019. Performance analysis of MAPbI3 based perovskite solar cells employing diverse charge selective contacts: Simulation study. Sol. Energy 193, 948–955.

[16] Shamna, M. S., Nithya, K. S., & Sudheer, K. S. (2020). Simulation and optimization of CH3NH3SNI3 based inverted perovskite solar cell with NIO as hole transport material. Materials Today: Proceedings, 33, 1246–1251.

[17] Minemoto, T., Kawano, Y., Nishimura, T., Shen, Q., Yoshino, K., Iikubo, S., Hayase, S., & Chantana, J. (2020). Theoretical Analysis of band alignment at Back Junction in sn– Ge perovskite solar cells with inverted P-i-N structure. Solar Energy Materials and Solar Cells, 206, 110268.

[18] Jalalian, D., Ghadimi, A., & Kiani, A. (2019). Modeling of a high performance bandgap graded PB-free HTM-free perovskite solar cell. The European Physical Journal Applied Physics, 87(1), 10101.
[19] Mandadapu, U. (2017). Simulation and analysis of lead based perovskite solar cell using SCAPS-1D. Indian Journal of Science and Technology, 10(1), 1–8.
[20] Islam, M. A., Wahab, Y. A., Khandaker, M. U., Alsubaie, A., Almalki, A. S., Bradley, D. A., & Amin, N. (2021). High mobility reactive sputtered CuxO thin film for highly efficient and s^{table} perovskite solar cells. Crystals, 11(4), 389.
[21] Eperon, G.E., Leijtens, T., Bush, K.A., Prasanna, R., Green, T., Wang, J.T.-W., McMeekin, D.P., Volonakis, G., Milot, R.L., May, R., 2016a. Perovskite-perovskite tandem photovoltaics with optimized band gaps. Science 354 (6314), 861–865.
[22] Singh, N., Agarwal, A., & Agarwal, M. (2020). Numerical simulation of highly efficient lead-free all-perovskite tandem solar cell. Solar Energy, 208, 399-410.
[23] Correa-Baena, J. P., Anaya, M., Lozano, G., Tress, W., Domanski, K., Saliba, M., ... & Hagfeldt, A. (2016). Unbroken perovskite: interplay of morphology, electro-optical properties, and ionic movement. Advanced Materials, 28(25), 5031-5037.
[24] Burdick, J., Glatfelter, T., 1986. Spectral response and I-V measurements of tandem amorphous-silicon alloy solar cells. Solar Cells 18 (3), 301–314.
[25] Chen, B., Yu, Z., Liu, K., Zheng, X., Liu, Y., Shi, J., ... & Huang, J. (2019). Grain engineering for perovskite/silicon monolithic tandem solar cells with efficiency of 25.4%. Joule, 3(1), 177-190.
[26] Pandey, R., Singla, A., Madan, J., Sharma, R., Chaujar, R., 2019. Toward the design of monolithic 23.1% efficient hysteresis and moisture free perovskite/c-Si HJ tandem solar cell: a numerical simulation study. J. Micromech. Microeng. 29 (6), 064001.
[27] Ramírez Quiroz, C. O., Spyropoulos, G. D., Salvador, M., Roch, L. M., Berlinghof, M., Darío Perea, J., ... & Brabec, C. J. (2019). Interface molecular engineering for laminated monolithic perovskite/silicon tandem solar cells with 80.4% fill factor. Advanced Functional Materials, 29(40), 1901476.
[28] Sahli, F., Werner, J., Kamino, B.A., Bräuninger, M., Monnard, R., Paviet-Salomon, B., Barraud, L., Ding, L., Diaz Leon, J.J., Sacchetto, D., Cattaneo, G., Despeisse, M., Boccard, M., Nicolay, S., Jeangros, Q., Niesen, B., Ballif, C., 2018a. Fully textured monolithic perovskite/silicon tandem solar cells with 25.2% power conversion efficiency. Nat. Mater. 17 (9), 820–826.
[29] Leijtens, T.; Eperon, G.E.; Pathak, S.; Abate, A.; Lee, M.M.; Snaith, H.J. Overcoming ultraviolet light instability of sensitized TiO2 with meso-superstructured organometal tri-halide perovskite solar cells. Nat. Commun. 2013, 4, 2885.
[30] Han, Y.; Meyer, S.; Dkhissi, Y.;Weber, K.; Pringle, J.M.; Bach, U.; Spiccia, L.; Chneg, Y.-B. Degradation observations of encapsulated planar CH3NH3PbI3 perovskite solar cells at high temperatures and humidity. J. Mater. Chem. A. R. Soc. Chem. 2015, 3, 8139–8147.
[31] Dong, L.; Sun, S.; Deng, Z.; Li,W.;Wei, F.; Qi, Y.; Li, J.; Li, X.; Lu, P.; Ramamurty, U. Elastic properties and thermal expansion of lead-free halide double perovskite Cs2AgBiBr. Comput. Mater. Sci. 2018, 141, 49–58.

IOP Conf. Series: Materials Science and Engineering 1278 (2023) 012006 doi:10.1088/1757-899X/1278/1/012006

Influence of solvent on accelerating deposition time of α-Fe_2O_3 photoanode via AACVD

Amin Aadenan[1,2,*], Nurul Affiqah Arzaee[1], Mohamad Firdaus Mohamad Noh[1], Mohd Norizam Md Daud[1], Inzamam Nawas Nawas Mumthas[1], Danial Hakim Badrul Hisham[1,3], Muhammad Athir Mohamed Anuar[1,3], Nurul Aida Mohamed[1], Mohd Adib Ibrahim[1], Norasikin Ahmad Ludin[1], Mohd Asri Mat Teridi[1,*]

[1]Solar Energy Research Institute, Universiti Kebangsaan Malaysia, 43600 Bangi, Selangor, Malaysia.
[2]Faculty of Applied Sciences, Universiti Teknologi MARA, Negeri Sembilan, 72000 Kuala Pilah, Negeri Sembilan, Malaysia.
[3]Renewable Energy & Green Technology, TNB Research, 43000, Kajang, Selangor, Malaysia.

*Corresponding Authors: asri@ukm.edu.my

Abstract. The optical and photoelectrochemical (PEC) properties of semiconductors can be manipulated by varying the deposition time and the solvent used. In this work, the study investigates the influence of organic solvent (methanol and ethanol) on accelerating deposition time for the efficient optical and photoelectrochemical performance of the fabricated α-Fe_2O_3 thin film photoanode below 10 minutes. The α-Fe_2O_3 film was fabricated via aerosol-assisted chemical vapor deposition (AACVD). The study discovered that the solvent with high polarity enhanced the crystal growth direction of α-Fe_2O_3 on the (110) orientation. The UV-vis showed that all deposited α-Fe_2O_3 films exhibit the absorption band in the visible-light region with the estimated energy bandgap of 2.10 eV. Furthermore, the PEC performance exhibited good photoanodic responses with excellent charge transfer behavior and photostability. These findings show that the choice of solvent could influence the rapid deposition time, thus, would be reliable to accelerate the fabrication process of photoelectrochemical water-splitting photoanode.

1. Introduction

Hematite (α-Fe_2O_3) is an *n*-type material with an optical bandgap energy range from 1.9 to 2.2 eV [1]. It has attracted considerable interest as a photoelectrode in PEC systems due to its excellent thermodynamic stability, non-toxicity, low cost, and abundance [2]. However, α-Fe_2O_3 experiences poor photoresponse due to high resistivity and charge recombination. Additionally, low charge mobility and short carrier diffusion length (2–4 nm) are other issues experienced by hematite which in turn limits its photoactivity to occur efficiently [3,4].

Various methods have been employed to synthesize α-Fe_2O_3 nanostructures including thermal oxidation [5], hydrothermal and solvothermal [6–8], plasma enhanced-chemical vapor deposition (PECVD) [9], and aerosol-assisted chemical vapor deposition (AACVD) [4,10]. Among of them,

AACVD is considered as an attractive option for the formation of nanostructured thin films due to facile process, easy to operate, versatile and most importantly easy to control the desired morphology film growth [11,12]. However, previous study showed that prolonging the AACVD deposition time could improve the light absorption [4], which in this case it is quite time-consuming for large-scale manufacturing. Additionally, the solvent used in dissolving the precursor for AACVD plays an important role in the reaction mechanism, providing that the precursor can react differently with various solvents in the gas phase [13]. Furthermore, studies have also demonstrated that the solvent used could enhance the growth direction of the crystal structure and influence the overall morphological formation of the samples [14,15]. Therefore, this will eventually result in the enhancement of optical and PEC performances of α-Fe_2O_3 photoanode.

In this work, we investigated the effect of organic solvent on accelerating the deposition time towards the optical and photoelectrochemical behaviors of α-Fe_2O_3 photoanode prepared by AACVD. The new idea of this work is to address that combining the two abovementioned factors will influence the modification of α-Fe_2O_3 photoanode behaviors. In this case, the study concentrates on optimizing the choice of organic solvent used in order to reduce the deposition time for fabricating the α-Fe_2O_3 thin film photoanodes. Therefore, the α-Fe_2O_3 films were fabricated by employing the AACVD with deposition time of 2, 5 and 10 minutes at 500 °C with the use of methanol and ethanol as the precursor organic solvents.

2. Experimental

2.1. Fabrication of α-Fe_2O_3 photoanodes by AACVD

Initially, fluorine-doped tin oxide (FTO) glass substrate (TEC 8, ~8Ω/sq Pilkington) was cut into (1 cm × 2 cm) and cleaned ultrasonically for 15 minutes in absolute ethanol (C_2H_6O 99.8% undenatured, Chemiz), 2-propanol ($CH_3CH(OH)CH_3$, EMSURE®) and acetone (C_3H_6O, Chemiz). The α-Fe_2O_3 photoanode was fabricated according to the previous published method [16]. In brief, 0.1 M precursor solution of $FeCl_3.6H_2O$ (97%, Sigma Aldrich) was dissolved in 50 mL methanol (CH_3OH, *Analytical Reagent* R&M Chemicals) and 50 mL absolute ethanol (C_2H_6O 99.8% undenatured, Chemiz), separately. The precursor was stirred at 550 rpm for 30 minutes until the mixture produces a clear solution with a bright orange (in methanol) and partially dark orange (in absolute ethanol), respectively. Next, the precursor solution was transferred into a 50 ml round-bottom flask that immersed in the piezoelectric modulator of an ultrasonic humidifier to generate an aerosol mist. Then, purified air was released to transport the vaporized precursor from the precursor chamber to the reaction chamber with the gas flow rate ratio of (1:2). In this work, the samples were subjected to fabrication for 2, 5 and 10 minutes of deposition time at 500 °C.

2.2. Characterization

X-ray diffraction (XRD) results were obtained by XRD analyzer (Bruker, D8 Advance) using Cu Kα radiation source with the scan range between 20° to 60°. Surface morphology was analyzed using environmental scanning electron microscope (ESEM) (FEI Quanta 450 FEG). Surface roughness was identified using atomic force microscopy (AFM) (Nanosurf Easyscan2). UV-Vis analysis was conducted using spectrometer (Lambda 35, PerkinElmer) to analyze the absorption spectra of films. Photoelectrochemical (PEC) measurement was performed by Potentiostat/galvanostat PGSTAT 204 with a three-electrode configuration system (Ag/AgCl reference electrode, Pt electrode, and the fabricated electrode) to measure the steady-state and chopped illumination current density on 1 cm^2 active area, performing in 0.5 M Na_2SO_4 (~pH 7) electrolyte under solar simulator (1 Sun illumination = 100 mW cm^{-2}). Electrochemical impedance spectroscopy (EIS) and chronoamperometry (CA) were performed using the similar Potentiostat/galvanostat PGSTAT 204 instrument.

3. Results and discussion

3.1. Influence of organic solvent on structural, morphological and surface roughness

Figure 1 (a) and (d) present the XRD patterns of α-Fe_2O_3 samples prepared using two different organic solvents methanol and ethanol. In general, the diffraction peaks confirmed that the XRD for both methanol and ethanol patterns are matched to the JCPDS file No. 00-002-0919 with a rhombohedral crystal structure of hematite. The strongest diffraction peak can be observed at $2\theta = 35.7°$ originated from (110) crystal plane for both samples. The other pronounced peaks can be indexed at $2\theta = 24.2°$, 33.3°, 41.0°, 49.5° and 64.0° with Miller indices (012), (104), (113), (024) and (300), respectively for sample prepared by using methanol solvent. However, the other pronounced peak for sample prepared using ethanol solvent can be indexed at $2\theta = 33.3°$ that corresponds to (104) crystal plane. Figure 1 (b) and (e) show the micro-scale SEM images of the prepared samples. It can be seen that the sample prepared using methanol forming the uneven flake-like with sharp apexes structure which could have more porous domain and slightly exposed the FTO substrate. This is quite different for a sample prepared using ethanol solvent where the film forming the irregular granule-like structure that the stacked granular particles can provide excessive diffusion paths, which later may slow the electron mobility. Figure 1 (c) and (f) portray the AFM topography of α-Fe_2O_3 prepared using methanol and ethanol, respectively. It can be shown that the film prepared by methanol possesses higher average surface roughness as compared to ethanol and was determined to be 136.18 nm and 134.80 nm, respectively. Therefore, it can be concluded that different organic solvents used to prepare the precursor solution could induce the variation of crystal growth and morphological formation of the α-Fe_2O_3 film. The variation in surface morphology might be due to the solvent used, as in this case, methanol is known as one of the solvents with the highest polarity of any organic solvent [14]. Thus, it would influence the overall performance of the fabricated devices. For the entire work, we opted to use methanol as solvent for AACVD in order to study the effect of deposition time on the optical and photoelectrochemical properties of the fabricated α-Fe_2O_3 photoanode.

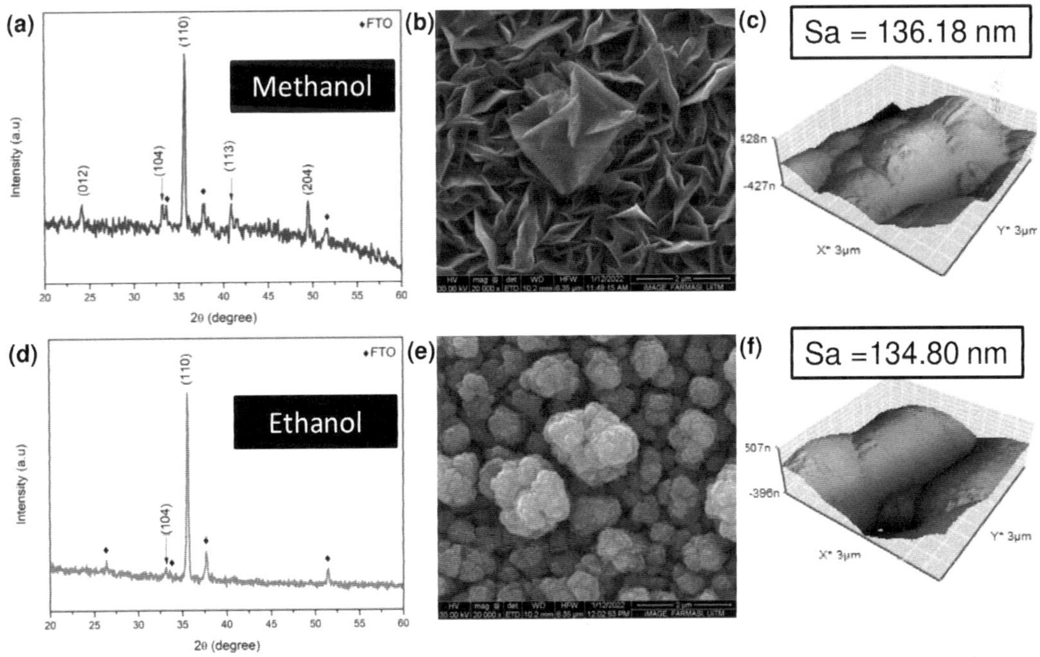

Figure 1. (a)–(c) XRD, SEM and AFM images of α-Fe_2O_3 prepared using methanol, and (d)–(f) XRD, SEM and AFM images of α-Fe_2O_3 prepared using ethanol solvent.

3.2. Optical properties

In this work, the optical properties of α-Fe_2O_3 films were analyzed using UV-vis absorption spectrophotometer. Figure 2 (a) demonstrates the absorbance spectra of α-Fe_2O_3 ranging from 380 – 700 nm deposited for 2, 5 and 10 min. In general, it is shown that the absorption curve of pristine α-Fe_2O_3 exhibits absorption edge within 580 – 600 nm in the visible light region and the absorption intensity becomes more intense as it reaches around 400 nm. In order to identify the optical bandgap energy (E_g) of α-Fe_2O_3, the Tauc plot was applied by extrapolating the absorption edge which derived by the following formula:

$$(\alpha h v)^n = A(hv - E_g) \qquad (1)$$

where α is absorption coefficient, hv is photon energy, n = 1/2 and 2 for direct and indirect band gap material, respectively, A is constant and E_g is bandgap energy. As can be seen in Figure 2 (b), by plotting $(\alpha h\upsilon)^{1/n}$ against hυ, it appears that the energy band gap of α-Fe_2O_3 all samples is 2.1 eV which in a good agreement with the previous literature [17]. Figure 2 (c) shows the Mott-Schottky plot of deposited α-Fe_2O_3 films that was carried out at 1 kHz frequency under dark conditions in 0.5 M Na_2SO_4 (~pH 7) electrolyte. Generally, the positive slopes from all deposited samples indicate that α-Fe_2O_3 films exhibited n-type characteristics. In order to determine the energy band edge positions of deposited films, the flat band potential (V_{fb}) can be obtained from the intersection between x-axis and the extrapolated line where the values are applicable for estimating the conduction band (CB) of α-Fe_2O_3. In this case, the flat band potentials obtained from 2, 5, and 10 min deposition time were 0.05, 0.01 and 0.03 V (vs. Ag/AgCl), respectively. The measured flat band potentials can be converted to the RHE scale using the Nernst equation:

$$V_{RHE} = V_{Ag/AgCl} + 0.209 + (0.059 \times pH) \qquad (2)$$

where $V^o_{Ag/AgCl}$ = 0.197 V (at 25 °C) and were recalculated as 0.67, 0.63 and 0.65 V (vs. RHE), respectively. Combining with the E_g obtained from the Tauc plot (Figure 2 (b)), the VB can be estimated using (CB = VB – E_g). Figure 2 (d) illustrates the estimated energy band diagram of α-Fe_2O_3 film deposited for 2, 5 and 10 min.

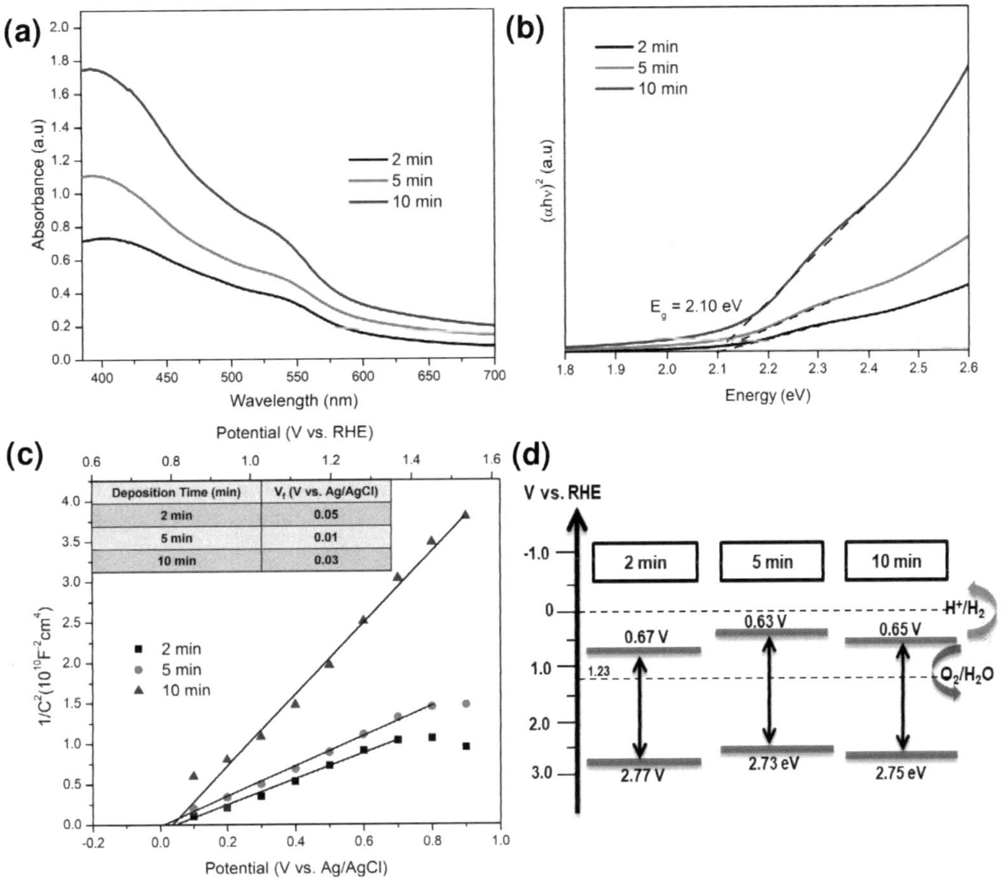

Figure 2. (a) Absorbance spectra, (b) Tauc plot, (c) Mott–Schottky plot, and (d) illustration of energy band diagram of α-Fe_2O_3 deposited at different time.

3.3. Photoelectrochemical (PEC) activity

The PEC performance of fabricated samples was tested to confirm their suitability for solar water splitting photoanode. Figure 3 (a) shows the current density-potential (J-V) characteristics of α-Fe_2O_3 films measured via linear sweep voltammetry (LSV) with the start potential begins at -0.2 V (vs Ag/AgCl) in 0.5 M Na_2SO_4 (~pH 7) electrolyte under 1.5 AM solar simulator. Overall, it can be seen that all the deposited α-Fe_2O_3 films by AACVD showing a good photoanodic (n-type) response both under dark and visible light illumination. The generated photocurrent onset for the three samples was recorded within 0.0 to 0.05 V (vs. Ag/AgCl) which the range is consistent with the increase in the flat band potential observed in the Mott–Schottky plot (as in Figure 2 (c)). The highest recorded photocurrent density at 1.23 V (vs. Ag/AgCl) is 0.85 mA/cm^2 for 5 min deposition, while the lowest reading is approximately 0.16 mA/cm^2 corresponds to 2 min sample. However, when prolonging the deposition time up to 10 min caused a slight decrease in the photo response to 0.66 mA/cm^2.

Figure 3 (b) demonstrates the photocurrent measured under light chopping conditions with the samples were front illuminated. The chopping patterns are comparable to the steady-state photocurrent response by showing good transient spike-like behavior at low potentials. Comparing the chopped light curve with the steady state J-V curve, all samples exhibit the transient spike-like behavior at low potentials. The light on/off transient responses at low potentials caused the spiking behavior corresponds to the recombination process involved between the interface of α-Fe_2O_3 and electrolyte.

IOP Conf. Series: Materials Science and Engineering 1278 (2023) 012006 doi:10.1088/1757-899X/1278/1/012006

These spikes are gradually diminished with increasing the potential and the photocurrent responses become more stepwise indicating that surface recombination is suppressed.

Figure 3 (c) portrays the Nyquist plot of α-Fe_2O_3 films measured by electrochemical impedance spectroscopy (EIS) with the frequency range from 0.01 Hz to 100 kHz under dark conditions. In EIS, it is known that the smaller the arc radius indicates the greater the charge transport efficiency. In this case, it can be observed that the arc radius of α-Fe_2O_3 films is arranged in the following order: 2 min > 10 min > 5 min. To discuss more thoroughly about the charge transfer of the α-Fe_2O_3 photoanodes, the electrochemical circuit fitting was performed by deploying the Randles equivalent circuit and the fitting data is summarized in the inset table (Figure 3 (c)). The smallest value of R_p (charge transfer resistance) for 5 min sample demonstrates the lowest charge transfer resistance between electrolyte and α-Fe_2O_3 surface. Conversely, the high R_p value for 2 min sample implies a slow charge carrier dynamic and therefore reduces the PEC performance. Overall, the quantitative results indicate that the EIS measurement is in good agreement with the current density-potential (J-V) characteristics with the 5 min sample presents the highest charge transfer and the most efficient separation of photogenerated charge carriers compared to other samples. Figure 3 (d) displays the short-term chronoamperometry (CA) test for 5 min deposition sample under steady illumination measured at 0.5 V (vs. Ag/AgCl). The photostability of this sample α-Fe_2O_3 shows relatively stable performance under constant illumination for a duration of 60 s with a very little sign of photocurrent degradation.

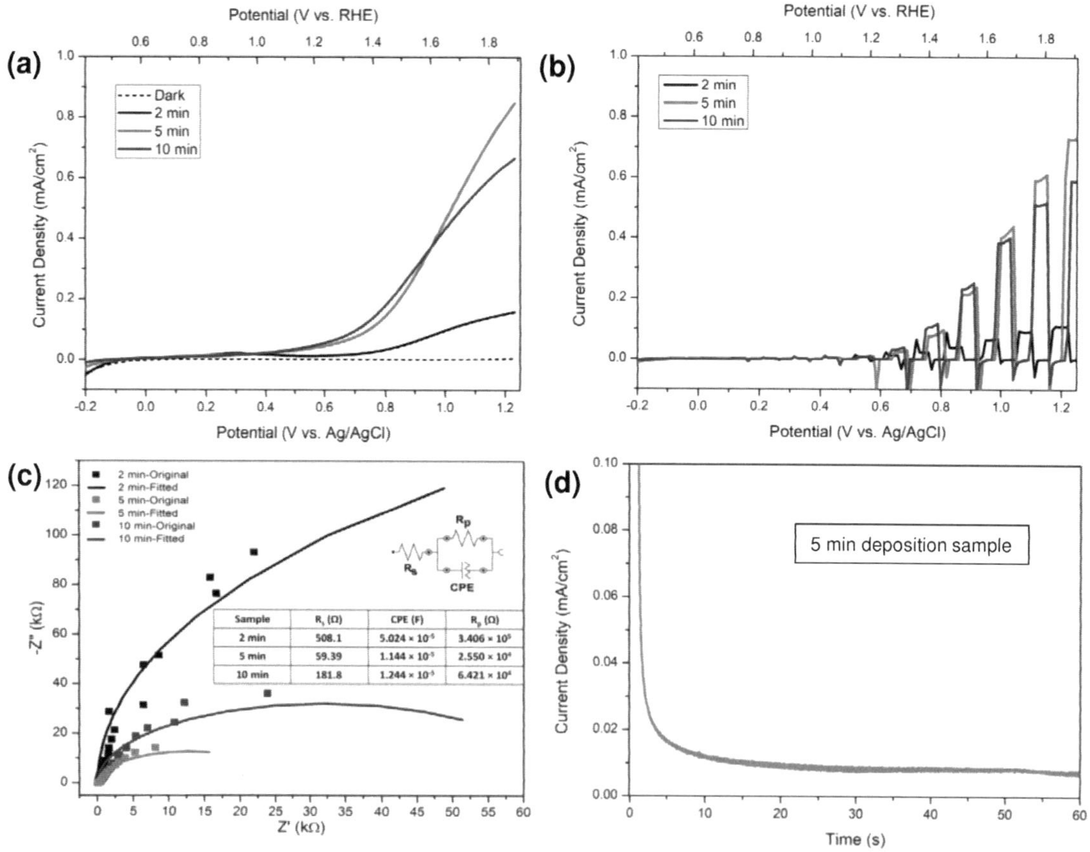

Figure 3. (a) Current density-potential (J-V) characteristics, (b) Photocurrent-responses under chopped illumination, (c) EIS Nyquist plot of α-Fe_2O_3, and (d) short duration Chronoamperometry (CA) test of 5 min deposition sample under constant illumination at 0.5 V (vs. Ag/AgCl) for 60 s in 0.5 M Na_2SO_4 (~pH 7) electrolyte.

4. Conclusion

In summary, α-Fe_2O_3 thin film photoanodes with different deposition time (2 min, 5 min, and 10 min) using methanol and ethanol as precursor solvents were successfully fabricated via AACVD with the best sample was recorded at 5 min deposition period. The results showed that the solvent with high polarity influenced the morphological formation of α-Fe_2O_3 crystal structure predominantly on the (110) crystal plane. In this case, the uneven flake-like structure was transformed into the granule-like structure by employing methanol and ethanol solvents, respectively. The enhanced surface morphology impacted on the improvement of optical function as well as the charge transportation of the fabricated photoanode. Moreover, the photocurrent density exhibited good photoanodic responses, resulting in the suppressed surface recombination, lower charge resistance and good stability. Hence, it is expected that the optical and photoelectron activities can be further enhanced by optimizing various parameters for fabricating α-Fe_2O_3 photoelectrodes for water-splitting applications.

Acknowledgement

The authors would like to thank Universiti Kebangsaan Malaysia (UKM) for the financial (GP-K016172-2020) and facility support, Ministry of Higher Education, Malaysia (MOHE) and Universiti Teknologi MARA (UiTM) for the SLAB scholarship.

References

[1] Cao J, Kako T, Kikugawa N and Ye J 2010 *J. Phys. D. Appl. Phys.* **43** 325101
[2] Al-Kuhaili M F 2012 *J. Alloys Compd.* **521** 178
[3] Tahir A A and Wijayantha K G U 2010 *J. Photochem. Photobiol. A Chem.* **216** 119
[4] Arzaee N A, Mohamad Noh M F, Ab Halim A, Abdul Rahim M A F, Mohamed N A, Safaei J, Aadenan A, Syed Nasir S N, Ismail A F and Mat Teridi M A 2019 *Ceram. Int.* **45** 16797
[5] Peng Q, Wang J, Feng Z, Du C, Wen Y, Shan B and Chen R 2017 *J. Phys. Chem.* C **121** 12991
[6] Mohammadikish M 2014 Ceram. Int. 40 1351
[7] Rajendran R, Yaakob Z, Mat Teridi M A, Abd Rahaman M S and Sopian K 2014 *Mater. Lett.* **133** 123
[8] Zheng M, Xing C, Zhang W, Cheng Z, Liu X and Zhang S 2020 *Inorg. Chem. Commun.* **119**
[9] Warwick M E A, Kaunisto K, Barreca D, Carraro G, Gasparotto A, Maccato C, Bontempi E, Sada C, Ruoko T P, Turner S and Van Tendeloo G 2015 *ACS Appl. Mater. Interfaces* **7** 8667
[10] Tahir A A 2014 *Phys. Status Solidi - Rapid Res. Lett.* **8** 976
[11] Marchand P, Hassan I A, Parkin I P and Carmalt C J 2013 *Dalt. Trans.* **42** 9406
[12] Mohamad Noh M F, Ullah H, Arzaee N A, Ab Halim A, Abdul Rahim M A F, Mohamed N A, Safaei J, Mohd Nasir S N F, Wang G and Mat Teridi M A 2020 *Dalt. Trans.* **49** 12037
[13] Edusi C, Sankar G and Parkin I P 2012 *Chem. Vap. Depos.* **18** 126
[14] Potter D B, Parkin I P and Carmalt C J 2018 *RSC Adv.* **8** 33164
[15] Kumar V, Singh K, Kumar A, Kumar M, Singh K, Vij A and Thakur A 2017 *Mater. Res. Bull.* **85** 202
[16] Aadenan A, Arzaee N A, Noh M F M, Mumthas I N N, Daud M N M, Mohamed N A and Teridi M A M 2022 *Bull. Mater. Sci.* **45**
[17] Bouhjar F, Ullah S, Chourou M L, Mollar M, Marí B and Bessaïs B 2017 *J. Electrochem. Soc.* **164** H936

IOP Conf. Series: Materials Science and Engineering 1278 (2023) 012007 doi:10.1088/1757-899X/1278/1/012007

Single-walled carbon nanotube (SWCNT) thin film for optoelectronics device prepared by spray coating technique.

Arulampalam Kunaraj[1], Iskandar Yahya[1*], Ahmad AA Bakar[1], Puvaneswaran Chelvanathan[2]

[1]Department of Electrical, Electronic & Systems Engineering, Universiti Kebangsaan Malaysia, Bangi, Selangor, 43600, MALAYSIA.

[2]Solar Energy Research Institute (SERI), Universiti Kebangsaan Malaysia, Bangi, Selangor, 43600, MALAYSIA.

E-mail: iskandar.yahya@ukm.edu.my

Abstract. CNT is a novel material with exceptional properties. SWCNT thin films have been used in many different applications. To fabricate the SWCNT film, various methodologies have been introduced, however, the problem of fabricating uniform SWCNT thin film remains unsolved. Spray coating technique (SCT) is the most effective method for producing a uniform SWCNT thin film. The CNT thin films were deposited on glass substrates via SCT with different nozzle distances (from 2.5 cm to 12.5 cm). The morphological, topological, electrical and optical properties of CNT thin films were characterized using appropriate techniques. The electrical analysis depicted that the sheet resistance of thin films reduced from 40220 Ω/sq. to 1330 Ω/sq. as the spray nozzle distance was increased from 2.5 cm to 10 cm sheet resistance reached 2133 Ω/sq. for 12.5 cm which confirmed the concentration of CNT was highest at 10 cm nozzle distance. The optical analysis of the CNT thin films gave transmittance values in percentage of 64%, 60%, 52%, 43%, and 46%, respectively for the nozzle distance of 2.5 cm, 5 cm, 7.5 cm, 10 cm, and 12.5 cm. Transmittance percentage decreased from 2.5 cm to 10 cm nozzle distance, indicative of an uniform film coverage of CNT layer. The findings of this study could be a platform for producing an optimized CNT thin film using a SCT for optoelectronic devices.

1. Introduction

Thin films are used extensively in modern technology [1].The demand for thin films has skyrocketed in the modern era, owing primarily to the rapid miniaturization of electronic devices. Thin films are now widely used in engineering and science technology to improve the performance of engineering systems and products. Thin film is defined as a layer that broadens in two directions but is limited in the third dimension. Thickness ranging from a few nanometres to a few micrometres. Because of their many advantages, such as good flexibility and low weight, nanostructured materials are frequently used in thin film production. Thin films are classified according to their, electrical properties, optical properties, chemical properties, thermal properties and mechanical properties.

Carbon nanotubes (CNTs) are a type of nanomaterial with exceptional optoelectronic [2][3] and mechanical properties [4]. Due to the hollow structure, large specific surface area, good electrical, thermal, and mechanical properties, carbon nanotubes (CNTs) have received a lot of attention [5][6], because of these properties CNTs have used in different type of application[4] include thin film transistors, solar cells, sensors, supercapacitors, interactive display panels. Besides that, CNTs are also commonly used as thin film materials, particularly in the optoelectronics application.

There are currently some significant challenges in producing nanostructured thin films such as cost-effective deposition method and fabricating uniform homogeneous thin film [7]. Thin-film deposition methods are classified into two types such as physical method and chemical method [1]. Physical vapour deposition, molecular beam epitaxy, laser ablation and sputtering are examples of physical methods. Chemical methods include gas-phase techniques and solution techniques. Chemical vapour deposition [8] and atomic layer epitaxy [9] are two type of gas-phase methods. spray pyrolysis [10] ,dip-coating [10], spin coating and, sol-gel are under the solution methods. However, the spray coating technique is an excellent chemical method to fabricate thin film. It is simple [11] and low-cost method for producing thin films [12].

This work describes the optical and electrical properties of a thin film of SWCNTs deposited using air brush spray coating technique. Electrical and optical characterization were carried out to study the correlation between spray nozzle distance to the thin film morphological and topological properties, hence establishing the optimal spraying parameters.

2. Experimental details

2.1 Preparation of SWCNTs solution

Times Nano Energy Tech Co., Ltd China provided the High Purity Single-walled Carbon Nanotubes (Purity: >95wt%). To create a stable solution, 10 mg SWCNT was mixed with 40 ml DI water and 2% TX-100(Sigma-Aldrich). TX-100 acts as a solvent to improve dispersion [13][14]. This solution mixture was sonicated with an ultrasonic homogenizer tip probe inside the ice bath. To prevent SWCNT breakdown, run the homogenizer for two minutes and then turn it off for one minute for six cycles.

2.2 Substrate preparation

All experimental works for fabrication thin films have been carried out in open air [15].For the substrates, 2.5 mm by 2.5 mm clean square glass slides were used. Glass substrates were brushed front and back side for five times and this process repeated one more time for front side of the glass. Following this process, brushed glass substrates cleaned with using DI water. These glass substrates kept in baker with methanol which covered using aluminium and this process kept for 10 minutes. Again, same previous methodology applied for ethanol for 5 minutes and methanol for 5 minutes. After sonification process using ethanol and methanol, both glass substrates and beaker cleaned using DI water. These glass substrates kept in baker with DI water covered aluminium and sonification for 10 minutes, and glass substrates cleaned using N_2 Gas. Following this process, glass substrates transferred to hotplate and kept at 80 °C [16]. Finally, Ozone treatment did for glass substrates around 10 minutes.

2.3 Experimental setup

Figure 1 depicts an overall view of the experiment setup that conducted inside the fume chamber. The glass substrate is placed inside the glass substrate holder, which is positioned on top of the hotplate and directly beneath the Artistic air brush. A glass substrate holder is used to keep the glass substrate in right position while spraying. Hotplate and air compressor are powered by a 230V AC supply.

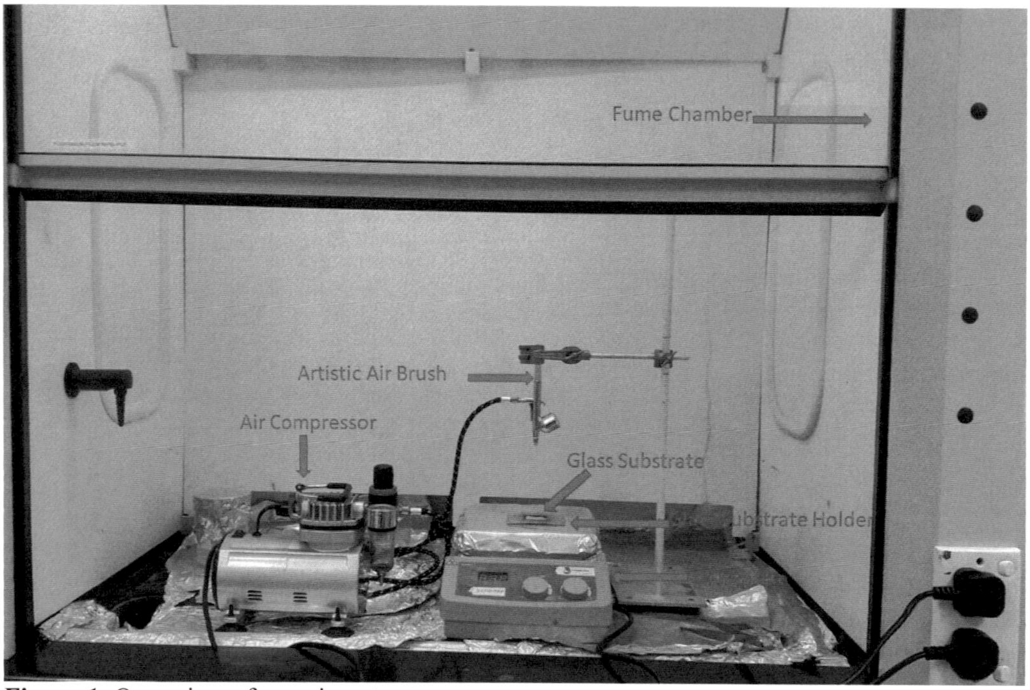

Figure 1. Over view of experiment setup

2.4 Deposition method of SWCNTs

The air compressor hose is first connected to the artistic airbrush hose connector and the pressure is set to 15 psi. An infrared thermometer is the tool used for the measurement of this surface temperature. The glass slide is then placed on the hot plate and directly under the nozzle. Due to air pressure, the length between the nozzle point and the glass substrate can have an effect on the SWCNTs network [17][18][19]. Finally, vary the distance between the substrate and the nozzle by 2.5 cm. 5 cm, 7.5 cm, 10 cm, and 12.5 cm Using an airbrush, spray SWCNTs for 0.5 seconds for each distance mentioned. After spraying SWCNT solution on glass substrate, immediately evaporate SWCNT solution with hotplate temperature 80 °C to avoid SWCNT bundling in the target area due to surface tension [20]. Using nozzle distances of 2.5 cm, 5 cm, 7.5 cm, 10 cm, and 12.5 cm, samples a, b, c, d, and e were sprayed.

3. Result and Discussion

3.1 Physical observation

Figure 2 shows physical observation of SWCNTs thin film sprayed at different nozzle distances. As can we seen clearly in figure SWCNTs film sprayed at 2.5 cm shows large non-uniform ring light film is formed whereby SWCNTs film only deposited at the edge, while centre ring uncovered. Where by samples a and b predominantly consists of glass surfaces in the centre of film. Films deposited at 7.5 cm and above shows better coverage of SWCNTs.

Figure 3 depicts the CNT thin film uniformity phenomenon at 2.5 cm and 10 cm nozzle distances. Because the initial force and CNT spray amount are equal, the initial velocity V is the same. In Figure 3(a), the CNT particles hit the glass substrate with more velocity, causing solution particles to splash away from the exact location, which is why a pin hole was created in the centre of the film. Figure

3(b) shows that CNT particles hit the glass substrate with moderate velocity, so there is less chance of splashing away from the exact position, which is why the thin film has good uniformity at 10 cm nozzle distance.

In the nozzle, the edge point is designed in the shape of a ring. The solution only comes out of the ring's edge. The inside of the ring area is completely covered, so no solutions can come out from inside the ring area. When the nozzle distance is small, there is less chance of spraying under the ring area in the glass substrate. When the nozzle distance is increased (10 cm), there is a greater chance of spraying all over the area under the nozzle. The nozzle ring is another reason for the uniform thin film at 10cm nozzle distance instead of 2.5 cm nozzle distance.

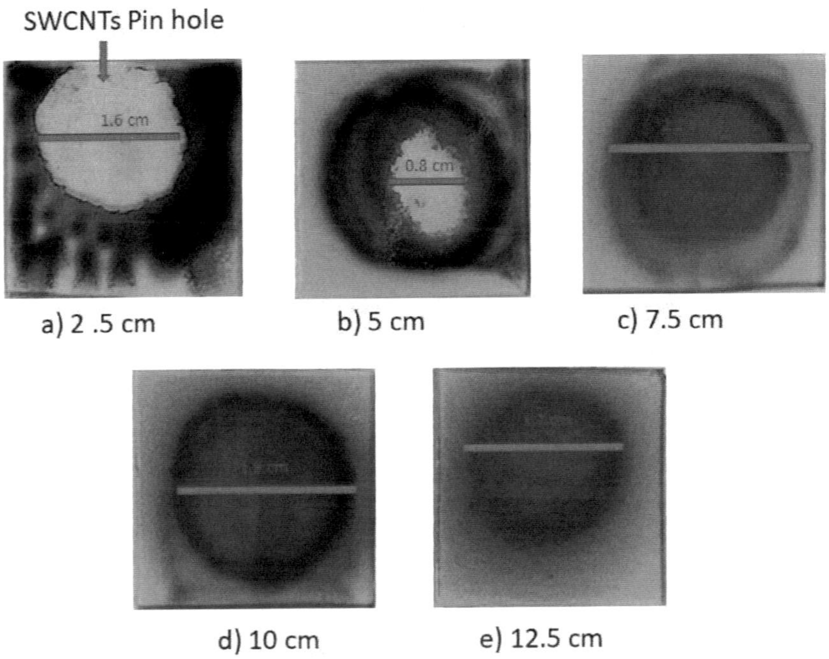

Figure 2. Physical images of SWCNT thin film samples sprayed at varying nozzle distance of a) 2.5 cm b) 5 cm c) 7.5 cm d) 10 cm e) 12.5 cm.

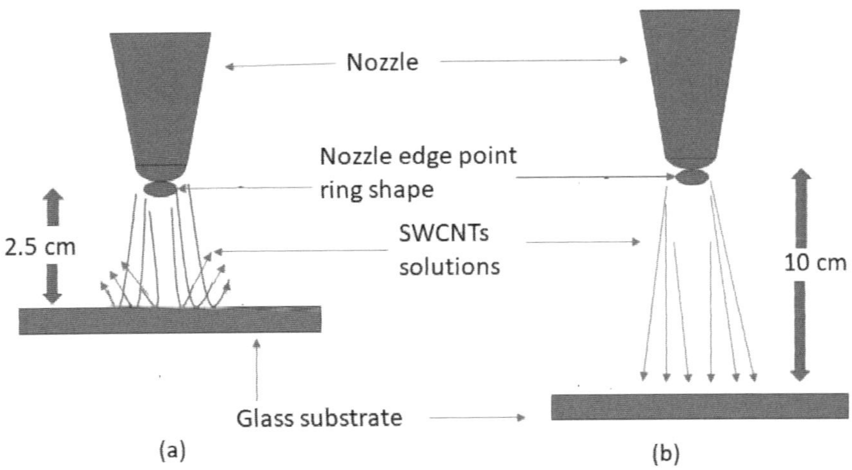

Figure 3. Illustration of the spray nozzle showing the SWCNT thin film spraying mechanism at two different nozzle distance of (a) 2.5 cm and (b)10 cm.

3.2 G1200 Digital Microscope image

Figure 4 shows high magnification microscopy images (1200 x magnification) of various nozzle distance samples. Because of the nozzle edge point ring shape and the shorter distance between the nozzle and samples, there is less chance of spraying SWNCT in the centre of the films (Figure 4(a) and Figure 4(b)). As a result, Figure 4(a) and Figure 4(b) have lower concentrations of SWCNT in the centre of the films. Figure 4(d) shows that SWNCT was sprayed very uniformly at a distance of 10 cm from the nozzle.

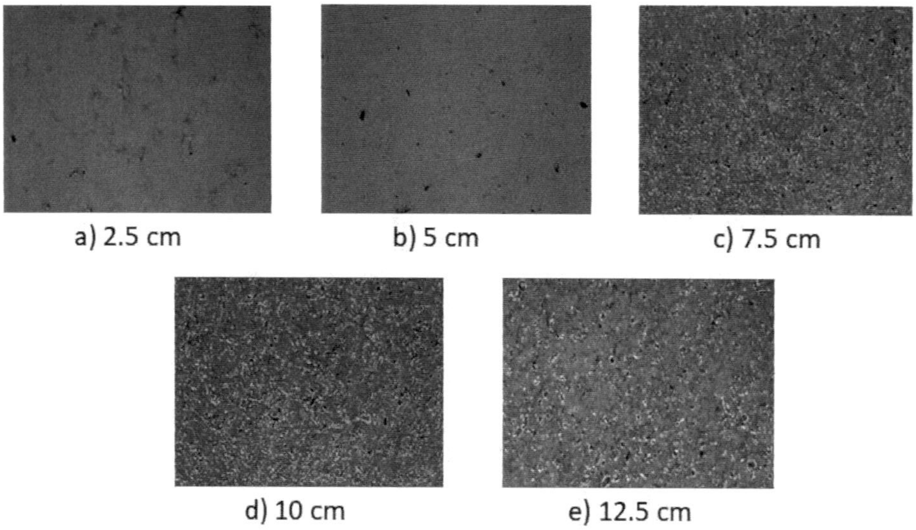

Figure 4. Microscope image (1200 x magnification) of SWCNT thin film surface sprayed at varying nozzle distances of a) 2.5 cm b) 5 cm c) 7.5 cm d) 10 cm e) 12.5 cm

3.3 4-point prob electrical characterises

The sheet resistance values of various nozzle distance samples are shown in Table 1. The electrical analysis depicted that the sheet resistance of thin films reduced from 40220 Ω/sq. to 1330 Ω/sq. As the spray nozzle distance was increased from 2.5 cm to 10 cm and then sheet resistance reached 2133 Ω/sq. for 12.5 cm nozzle distance, which confirmed the SWCNT uniformity was the highest at 10 cm nozzle distance.

Table 1. Sheet Resistances of SWCNT thin film samples sprayed at varying nozzle distance of a) 2.5 cm b) 5 cm c) 7.5 cm d) 10 cm e) 12.5 cm.

Sample	Sheet resistance(Ω/sq.)
a	40220 Ω/sq.
b	2770 Ω/sq.
c	1960 Ω/sq.
d	1330 Ω/sq.
e	2133 Ω/sq.

The resistivity can be easily calculated from the below equation using the sheet resistance value.

$$R_S = \frac{\rho}{t} \quad \ldots\ldots\ldots\ldots\ldots\ldots \text{(1.1)}$$

Where
Rs =sheet resistance
ρ=resistivity

3.4 *UV-Vis-spectrophotometer Transmission spectra of CNT thin films*

Figure 5 shows the transmittance values of various nozzle distance samples. The optical analysis of the CNT thin films gave transmittance values in percentage of 64%, 60%, 52%, 43%, and 46%, respectively for the nozzle distance of 2.5 cm, 5 cm, 7.5 cm, 10 cm, and 12.5 cm at 1400nm wavelength. Transmittance percentage decreased from 2.5 cm to 10 cm, indicative of an uniform film coverage of CNT layer. Because SWCNT concentration is low at 2.5 cm nozzle distance, the transmittance percentage is high. Simultaneously, at nozzle distance of 10 cm, SWCNT concentration is high, resulting in a low transmittance percentage. At 900 nm wavelength, there is a sudden jump due to the CNT pin holes in the nozzle distances of 2.5 cm and 5 cm. By obtaining the transmittance values of the films, we can calculate the absorbance values using the following equations.

$T = \frac{I}{I_0}$.................. (2.1)

$A = \log\frac{I}{I_0}$............ (2.2)

Where
I= Light intensity after passing through the sample.
I_o=Intensity of light before passing through the sample.
A=Absorbance.

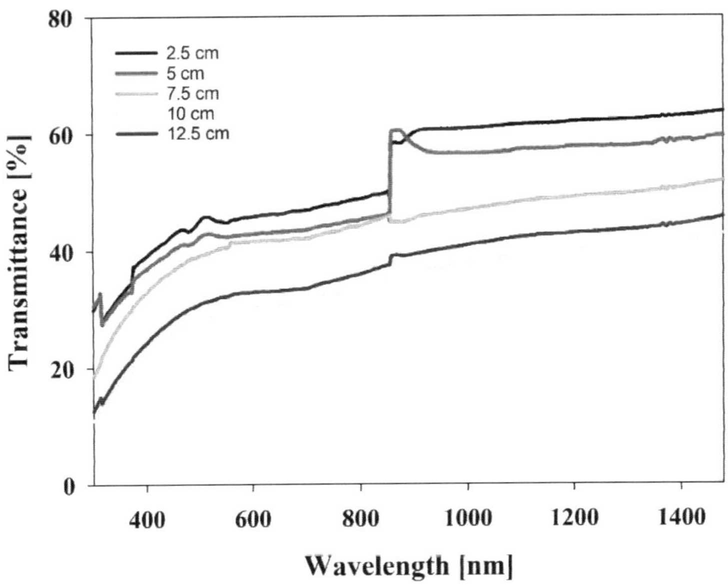

Figure 5. UV-Vis spectra showing the Transmittance vs Wavelength graphs for SWCNT thin films sprayed at varying nozzle distances of a) 2.5 cm b) 5 cm c) 7.5 cm d)10 cm e) 12.5 cm

3.5 Optimum sample

Result obtained from physical image, microscopy image and 4-point probe sheet resistances values, which was a 10 cm nozzle distance sample. The camera and microscope images clearly show an optimum sample of SWCNT dispersed well in glass substrate. 4-point analysis shows that the optimum sample had a low sheet resistance when compared to the other samples, which is 1330 Ω/sq. The transmission characteristics of the optimum sample had been low in comparison to other samples, but for thin film low sheet resistance is very important, so the 10 cm nozzle distance sample was chosen as the optimum sample.

3.6 FESEM Morphology characteristics

The FESEM morphology analysis was carried out in order to observe SWCNT network that forms the thin film. Figure 6 depicts the SWCNT network shape of an optimization (10 cm nozzle distance) sample at x250, x300, x500, and x30000 resolutions. In comparison to Figure6(a), Figure6(b) and Figure6(c) shows SWCNT dispersed well. Due to a lack of evaporation of SWNCT bundles from the hotplate and surface tension, some SWCNT bundles could be seen in Figure(c). To overcome this problem, the sample must be dried on a hotplate for more than 5 minutes [14].

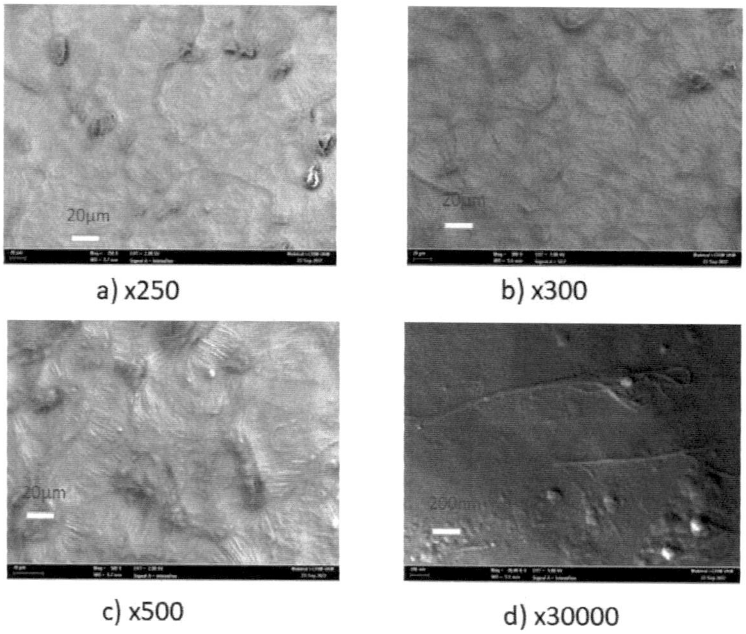

Figure 6. FESEM images of 10 cm nozzle distance SWCNT thin film resolution at a) x250 b) x300 c) x500 d) x30000.

3.7 AFM Topology characteristics

Figure 7 display AFM topology image of a 10 cm nozzle distance sample revealed that SWCNTs were uniformly coated over the glass substrate, with high concentrations of SWCNTs in the mid and edge areas. Only a few sample locations were missed in order to disperse SWCNT well. According to the AFM topology image, the film average roughness is 3.3 nm.

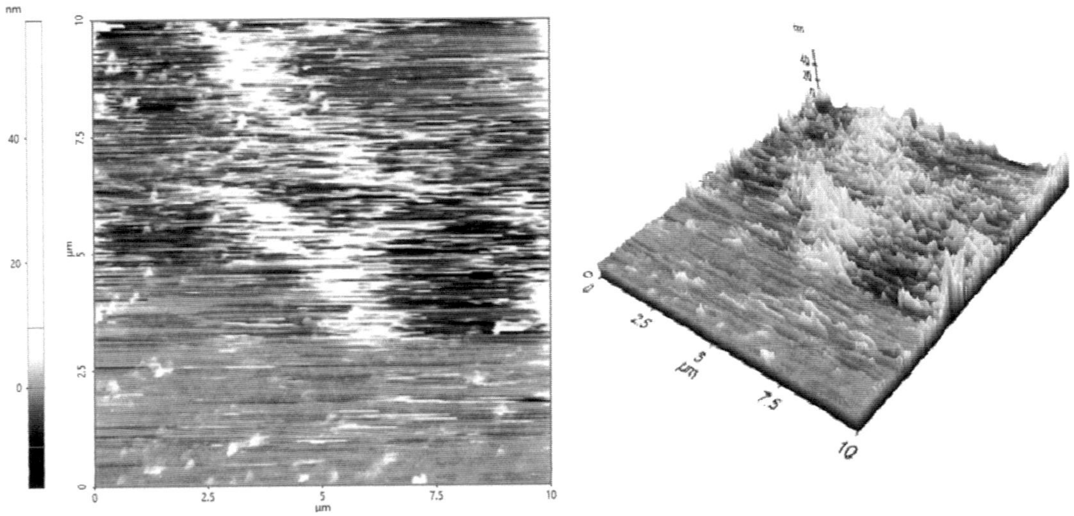

(a) (b)

Figure 7. AFM images of 10 cm nozzle distance SWCNT thin film (a) 2D (b) 3D

4. Conclusion

Spray coating technology was used to create a thin film of SWCNT. The nozzle distance is an essential factor in the production of SWCNT thin films. The nozzle distance of 10 cm was identified as the optimum sample with a low sheet resistance of 1330 Ω/sq. SWCNTs dispersed well in glass substrate, as shown by this sample microscope image and AFM image. This research experiment completed successfully but there is still room for improvement in order to create more reproducible optoelectronics device.

Funding This research was funded by the Ministry of Higher Education of Malaysia through the Fundamental Research Grant Scheme (FRGS/1/2021/TK0/UKM/02/29) and the Research University Grant (GUP-2022-006) of Universiti Kebangsaan Malaysia (UKM).

Acknowledgments The authors would like to thank the Government of Malaysia through The Ministry of Higher Education, Malaysia for funding this project and to Universiti Kebangsaan Malaysia (UKM) for providing the support and required research facility; in particular at the Photonics Lab, Faculty of Engineering and Built Environment, Thin Film Solar Cell Laboratory in the Solar energy Research Institute (SERI).

References

[1] Perednis D and Gauckler L J 2005 Thin Film Deposition Using Spray Pyrolysis 103–11
[2] Yahya I, Theng L L, Mustaza S M, Abdullah H and Amin N 2017 Characterization of transparent conducting carbon nanotube thin films prepared via different methods *Sains Malaysiana* **46** 1103–9
[3] Dabera G D M R, Jayawardena K D G I and Prabhath M R R Hybrid Carbon Nanotube Networks as Efficient Hole Extraction Layers for Organic Photovoltaics 1–29
[4] John J C, Sebastian T, Cyriac J, Paul A, Jose A, Shaji S and Augustine S 2021 Optoelectronic Characteristics of In 2 S 3 -CNT Nanocomposite Thin Films for Photodetector Application *J. Electron. Mater.* **50** 2800–12
[5] Weng B, Yang M and Xu Y 2014 Toward the enhanced photoactivity and photostability of ZnO nanospheres via intimate surface coating with reduced graphene oxide † 9380–9

[6] Cervini R, Simon G P, Ginic-markovic M, Matisons J G, Huynh C and Hawkins S Aligned silane-treated MWCNT / liquid crystal polymer films **175602**

[7] Hecht D and Grüner G 2009 *Solution Cast Films of Carbon Nanotubes for Transparent Conductors and Thin Film Transistors*

[8] Ghoshtagore R N 1978 Mechanism of CVD Thin Film SnO Formation **1664** 110–7

[9] Suntola T 1992 Atomic layer epitaxy **216** 84–9

[10] I Yahya,SM Mustaza,H Abdullah 2018 Carbon Nanotube-Activated Thin Thin Film Transparent Conductor Applications.

[11] Tang C J, Azuddin Hassan M and Yahya I 2021 Direct Injection Carbon Nanotubes Thin Film Deposition System *19th IEEE Student Conf. Res. Dev. Sustain. Eng. Technol. Towar. Ind. Revolution, SCOReD 2021* 416–21

[12] SINGH V K, NO, SINGH V K and NO 2017 Thin Film Deposition By Spray Pyrolysis Techniques *Jetir* **4** 910–8

[13] Hassan M A, Mohamed M A, Kunaraj A and Yahya I 2022 Morphology and Electrical Characteristics of Single-Walled Carbon Nanotubes Film Prepared by Air Brush Technique *Int. J. Integr. Eng.* **14** 273–9

[14] Yahya I, Hassan M A, Nasyifa N, Maidin M and Mohamed M A 2022 SWCNT Network - FET Device for Human Serum Albumin Detection

[15] Oyeshola H O, Adisa M A, Adejumo B K, Babalola K K, Agboluaje B A, Adedokun O and Sanusi Y K 2020 Effect of low temperature synthesis of carbon nanotube nanocomposite on the photovoltaic performance of anode buffer layer in polymer solar cells *IOP Conf. Ser. Mater. Sci. Eng.* **805**

[16] Abdelhalim A, Abdellah A, Scarpa G and Lugli P 2013 Fabrication of carbon nanotube thin films on flexible substrates by spray deposition and transfer printing *Carbon N. Y.* **61** 72–9

[17] Stulik J, Kroupa M, Blecha T and Hamacek A 2015 Carbon Nanotube dispersion preparation and deposition of thin layers for gas sensors *Proc. Int. Spring Semin. Electron. Technol.* **2015-Septe** 1–4

[18] Abdellah A, Abdelhalim A, Horn M, Scarpa G and Lugli P 2013 Scalable spray deposition process for high-performance carbon nanotube gas sensors *IEEE Trans. Nanotechnol.* **12** 174–81

[19] Pham N P, Burghartz J N and Sarro P M 2005 Spray coating of photoresist for pattern transfer on high topography surfaces *J. Micromechanics Microengineering* **15** 691–7

[20] Hassan M A, Mohamed M A, Abdullah H and Yahya I 2021 Effect of channel length on single walled carbon nanotubes thin film characteristics deposited via spray coating technique *Proc. - 2021 IEEE Reg. Symp. Micro Nanoelectron. RSM 2021* 169–72

IOP Conf. Series: Materials Science and Engineering 1278 (2023) 012008 doi:10.1088/1757-899X/1278/1/012008

Numerical Evaluation of Thermal Performance of Two-Phased Closed Thermosyphon for Solar Applications

Muhamad Fadhli Ramlee[1], Adnan Ibrahim[1], Hasila Jarimi[1], Noorliyana Ramlee[2], Ahmad Fazlizan[1*]

[1]Solar Energy Research Institute (SERI), Universiti Kebangsaan Malaysia, 43600 Bangi, Selangor, Malaysia.

[2]Department of Landscape Architecture, Faculty of Architecture and Ekistics, Universiti Malaysia Kelantan, 16300 Bachok, Kelantan, Malaysia.

*Corresponding Author: a.fazlizan@ukm.edu.my

Abstract. A heat pipe is an effective passive heat transfer device with a highly efficient heat transfer rate and thermal conductivity through evaporating and condensing a working fluid circulating in a vacuum-sealed container. For solar applications, heat pipes are usually used in evacuated tube collectors (ETCs). However, the use of heat pipes in solar photovoltaic/thermal (PVT) collector is limited. Generally, a heat pipe composes of three sections: the evaporator section, the condenser section, and the adiabatic section. A wickless heat pipe, also known as a closed thermosyphon or gravitationally assisted heat pipe, relies on gravity to return the working fluid to the evaporator. Various parameters could greatly influence the thermal performance of wickless heat pipes such as initial volume filling ratio (VFR), inclination angle, inner diameter of the heat pipes, ratio of evaporator/diameter and type of working fluid used. In this study, 2D transient computational fluid dynamics (CFD) simulations were performed using ANSYS Fluent to simulate the two-phase flow (water/steam) and to study the effect of different parameters on the heat transfer processes of a two-phase closed thermosyphon (TPCT) to increase its thermal performance. The 2D CFD simulations using multiphase volume of fluid (VOF) with two Eulerian phases could replicate the mass and heat transfer processes in comparison with published experimental data. A good agreement can be observed between the present CFD study with the published data in comparing the temperature profiles and thermal performance of the heat pipe. The numerical simulation of the evaporation and condensation process indicated that the thermal performance and characteristics of the TPCT are influenced by certain key parameters. In the future, full-scale 3D CFD simulations are planned to be performed to evaluate the thermal performance of heat pipes for solar photovoltaic/thermal (PVT) applications.

1. Introduction

Two-phase closed thermosyphons (TPCTs) have been widely used in a vast range of engineering applications that involve mass and heat transfer processes such as solar collectors, air conditioning (HVAC) systems, electronics thermal management and heat exchangers. This is attributed to their simple passive structure, good compactness, high thermal efficiency, and excellent reversibility [1]. Generally, TPCTs consist of 3 sections, the evaporator section at the bottom end where heat flux is added to vaporize the working liquid inside the heat pipe, the condenser section at the upper end where heat is released and the liquid vapour is condensed, and the adiabatic section in between the two sections, as depicted in Figure 1.

The thermal performance of a TPCT is affected by various factors such as the initial volume filling ratio (VFR), inclination angle, inner diameter of the TPCT, ratio of evaporator/diameter and type of working fluid used. Previous studies have been conducted on the effect of the above factors on the thermal performance of the TPCT. For instance, an experimental study has been performed by Robinson et al. [2] to investigate the influence low fill ratio on the heat and mass transfer mechanism of a thermosyphon with a relatively long evaporator length of 200 mm and small diameter (8mm). The results indicated that that the heat and mass transfer process was dominated by Geyser-type boiling at lower heat fluxes with associated low heat transfer coefficients in the evaporator and condenser sections. Jafari et al. [3] provided an extensive review of the state-of-the-art of TPCTs technology and recent published theoretical and experimental TPCTs studies including descriptions of the operating principle of TPCTs, the performance characteristics and the heat transfer analysis at the evaporator and condenser section of TPCTs. This can be used to gain a better insight into the mass and heat transfer process of TPCT and their renewable energy applications. Kim & Moon [4] utilized CFD simulation to visualize the two-phase flow and heat transfer of a TPCT of which the visualization of the flow process cannot be done experimentally. They also examined the thermal resistance tendencies for different fill ratio, heat pipe diameter, and heating power. As a result, the thermal resistance decreases when the water amount, heat pipe diameter, and heater power increase, indicating that temperature is quickly cooled down by latent heat.

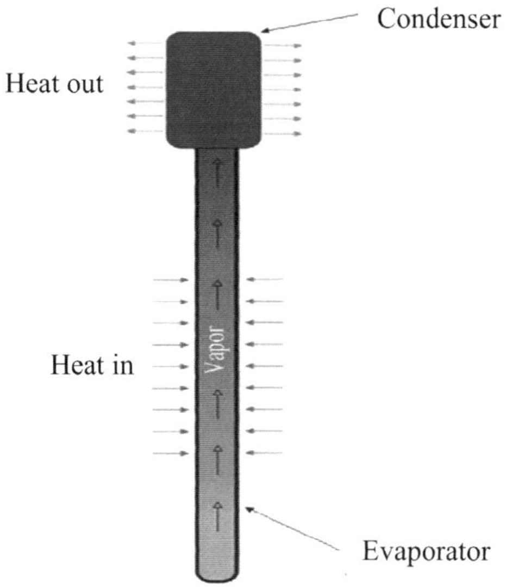

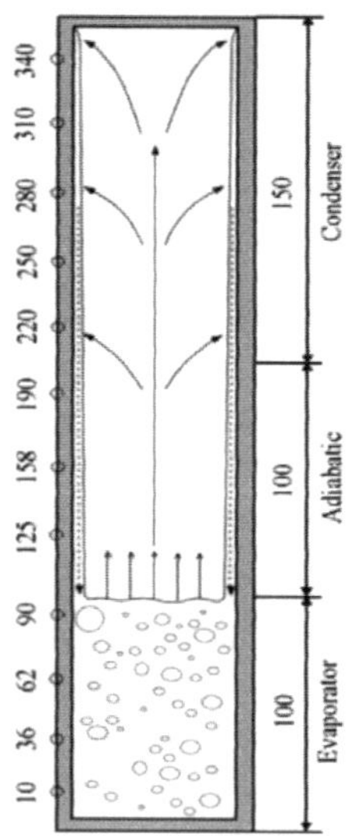

Figure 1. Schematic of a heat pipe [5]

The use of CFD technology in investigating the effect of various parameters on the performance TPCTs have been carried out by several researchers. For instance, Fadhl et al. [6] developed comprehensive CFD modeling using ANSYS software to simulate the heat transfer phenomena inside the TPCT i.e., condensation and evaporation processes. CFD results were validated against experimental data in comparing the temperature distribution along the heat pipe and thermal resistance at different heat inputs. They reported that the thermal performance of TPCTs improved by increasing heat input over 172 W. A steady state 3D CFD model has been developed by Anto & Varghese [7] to analyze heat transfer capability and the wall temperature distribution along the lateral walls of thermosyphon. The model was validated against experimental results. The CFD simulation helped to visualize the steady state wall temperatures over the entire wall of a TPCT at initial stage of research which is very difficult to be done experimentally.

In the present study, CFD simulation was carried out to investigate the influence of different working fluids and inclination angles on the thermal performance of a TPCT. A two-dimensional (2D) CFD model of copper TPCT is simulated with different working fluids (water, ethanol and acetone) at different inclination angles (90°, 45° and 20°) with different heating powers of 100.14W and 275.6W. The use of CFD models can reduce the amount of experimental work CFD as well as provide an insight of the complex flow phenomena inside a TPCT.

2. Methodology

2.1. Simulation Model and Mesh Strategy

A 2D model of TPCT was designed using Design Modeler software (ANSYS 2020 R1) with a 12mm diameter and 0.1m-long of copper tube. It consists of a 0.04m-long evaporator section, a 0.02m-long adiabatic section and a 0.04m-long condenser section, as shown in Figure 2(a). The ratio of evaporator, adiabatic and condenser sections are based on a previous study by [6] as the main reference in this study to perform CFD simulations. The detail of the parameters used in this study is listed in Table 1.

Table 1. Parameters of the TPCT

Parameters	Value
Length [m]	0.1
Diameter [m]	0.01
Material [-]	Copper
Heating Power [W]	100.41, 275.6
Working Fluid	Water, Ethanol, Acetone
Inclination Angle [°]	90, 45, 20

The computational grid is made up of structured cells (quadrilateral) considering its better accuracy compared to unstructured mesh. At the walls of the TPCT model, 10 layers of cells are set to capture the thin liquid film that develops near the wall. The details of the mesh size and type are shown in Fig. 2(b,c). The temperature along the TPCT was recorded at five different positions to monitor the temperature distribution, in which T_1 and T_2 were used to record the average temperature of the evaporator section, $T_{eva,\ avg}$, T_3 was used to record the temperature of the adiabatic section and T_4 and T_5 were used to record the average temperature of the condenser section, $T_{cond,\ avg}$.

2.2. Boundary Conditions and Computational Setup

A no-slip boundary condition is imposed at the walls of the TPCT. A constant heat flux is set at the evaporator's wall to simulate the heating and evaporation while a temperature of 300K is defined at the condenser's wall as the heat sink. A zero heat flux is defined as boundary condition on the adiabatic section. A combination of the SIMPLE algorithm for pressure-velocity coupling and a first-order upwind scheme for the determination of momentum and energy is included in the model. Geo-Reconstruct and PRESTO discretization for the volume fraction and pressure interpolation scheme, respectively, are also performed in the simulation

A transient simulation with the time-step size of 0.0003s is performed to simulate the dynamic behaviour of the heat transfer process in the TPCT. The time-step size is set based on the Courant number. For VOF models, the maximum Courant number allowed near the interface is 250 [8]. To ensure that the solutions achieve convergence, the simulation was set to run for a total number of time-steps of 200,000 with 20 iterations for each time-step. The simulation is considered to have converged when it reaches to a steady-state while achieving low scaled residuals of mass and velocity components ($<1 \times 10^{-4}$).

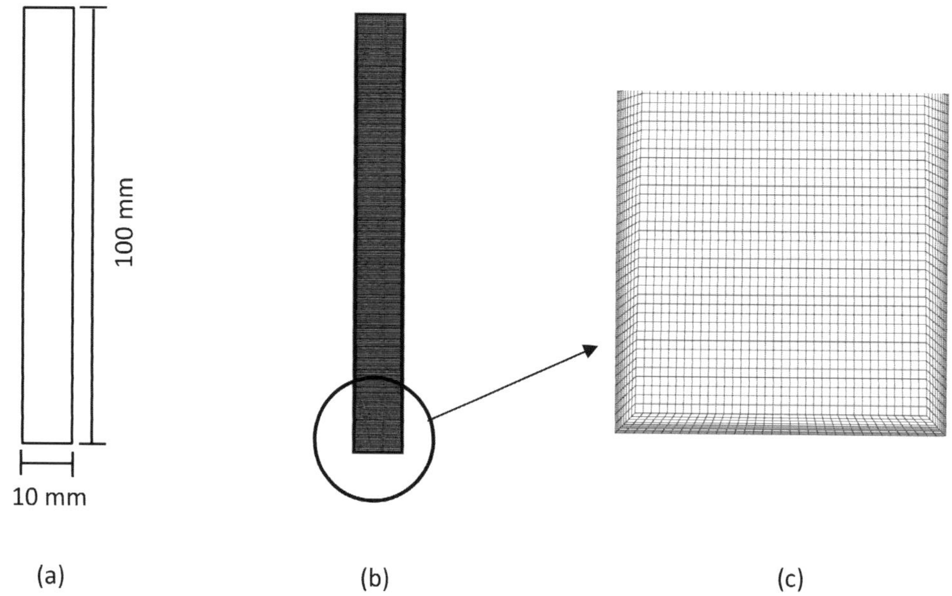

Figure 2. (a) Geometry of the 2D CFD model, (b) mesh of the model and (c) details of the cell size and inflation layers.

3. Simulation Validation and Results

3.1. Mesh Independence Study

To ensure the independency of the numerical results from grid mesh resolution, a basic mesh independence test has been carried out using three different mesh sizes i.e. coarse (10,765 cells), medium (18,020 cells) and fine mesh (25,368 cells). The average temperature of the evaporator, $T_{eva,\ avg}$ and condenser sections, $T_{cond,\ avg}$ for different mesh sizes were monitored, as shown in Table 2. For the heat input of 100.41 W, it was found that the differences between $T_{eva,\ avg}$ and $T_{cond,\ avg}$ of the medium and fine mesh are insignificant (1.09%) than those between the coarse and medium (6.22%). A notable difference in $T_{eva,\ avg}$ and $T_{cond,\ avg}$ can be observed between the coarse and medium grid, where coarse grid tends to over-predict the average temperature. Based on this grid independence analysis, the medium mesh size is found to produce a sufficiently accurate result with reasonable cost and time. Therefore, the medium grid is used for the rest of the following simulation in this paper.

Table 2. Mesh independence test

No. of cells	10,765 (coarse)	18,020 (medium)	25,368 (fine)
Average temp. of evaporator [K]	399.73	375.46	379.03
Average temp. of condenser [K]	340.81	321.62	322.64
Avg. percentage difference [%]	-	6.22	1.09

3.2. Validation of the CFD solution

To validate the CFD simulation, the same model geometry and operating conditions as [6] have been adopted. The temperature distribution at eight different positions along the TPCT for two different heat inputs (172.87W and 376.14W) have been compared with experimental work and CFD simulation by [6], as shown in Figure 3. The simulation results showed reasonable agreement with published experimental and numerical data with average percentage difference of less than 10%, considering our limited computational resources.

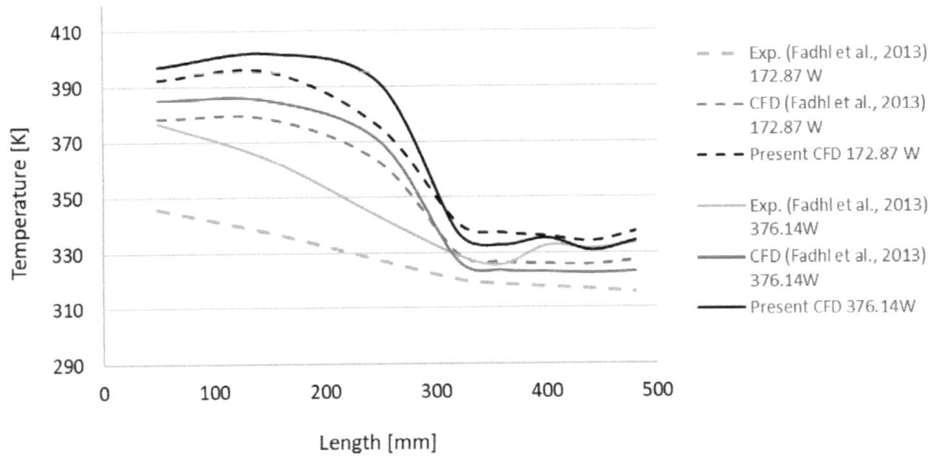

Figure 3. Comparison between published data by [6] and present CFD for heat input of 172.87W and 376.14W

3.3. Visualization of Evaporation and Condensation Process

Figure 4 shows the volume fraction contours of the evaporation process taking place in the evaporator and the condensation process in the condenser, respectively. A red colour illustrates the presence of only vapour (vapour volume fraction = 1), while a blue colour exhibits the presence of liquid only (vapour volume fraction = 0). Based on the figure, focus is made on the boiling process of the working fluid at the evaporator and the formation of condensed liquid film at the wall of the condenser. The initial process (t = 0) starts with the heating of the liquid pool that filled half of the evaporator by imposing a constant heat input, $\dot{Q}$. At positions where the liquid reached the saturation temperature (373.15 K), the liquid starts to evaporate and phase change occurs, as shown in Figure 4(b). This continuous evaporation of liquid results in a decrease in the liquid volume fraction and an increase in the vapour volume fraction. At those positions where the liquid evaporates, bubbles are formed and transported toward the top region of the liquid pool. Following the evaporation process, the saturated vapour is transported upward to the condenser. As the vapour reaches the condenser's wall, the vapour condenses along the cold walls forming thin liquid film (Figure 4(a)). This liquid will then fall back to the evaporator section and recharge the liquid pool. Similar flow phenomena can be found in the study by [9]and [10].

3.4. Performance of the TPCT

The performance of a TPCT can be evaluated by the overall thermal resistance. The overall rate of heat transfer to the system, $\dot{Q}$ is proportional to the temperature difference, ΔT between the heat source (evaporator) and the heat sink (condenser), and inversely proportional to the equivalent thermal resistance, R_{th} between the two regions, as shown in Eq. (1).

$$\dot{Q} = \frac{\Delta T}{R_{th}} \tag{1}$$

Therefore, the overall thermal resistance of a TPCT, R_{th} can be calculated using the Eq. (2):

$$R_{th} = \frac{T_{eva,avg} - T_{cond,avg}}{Q_{in}} \tag{2}$$

where $T_{eva,avg}$ and $T_{cond,avg}$ are the measured average temperature in the evaporator and condenser section, respectively, while Q_{in} is the heating power input.

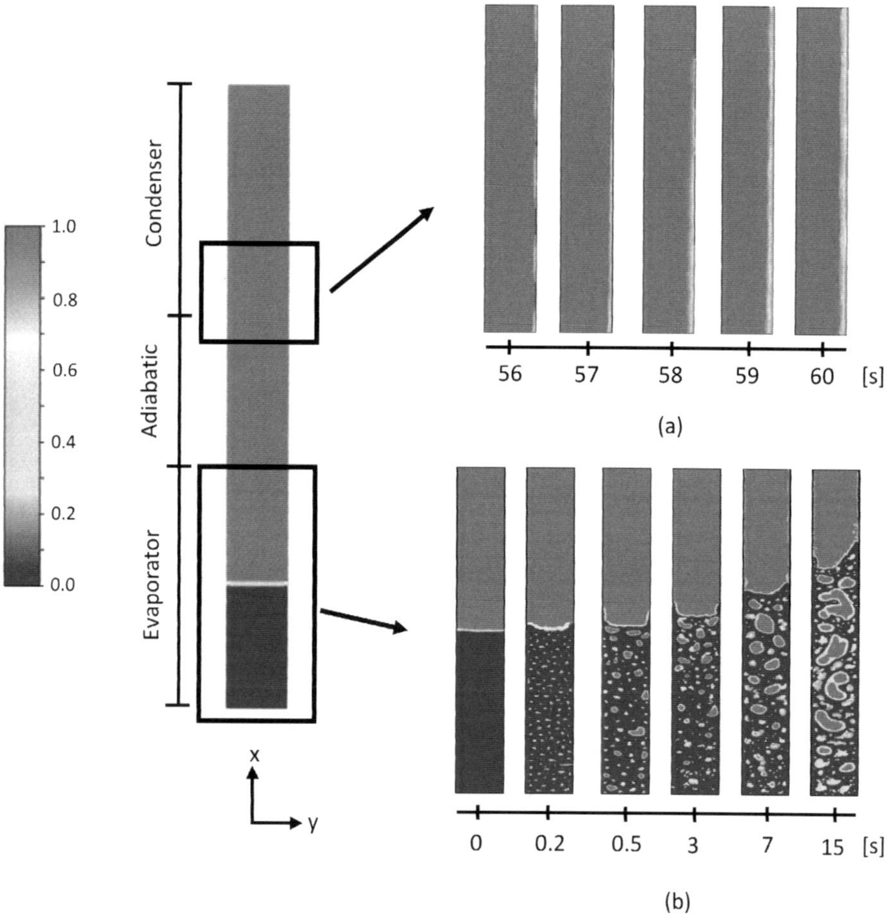

Figure 4. Volume fraction contour of (a) condensed liquid film at the condenser wall and (b) pool boiling in the evaporator section at different times.

Figure 5(a-b) shows the comparison of temperature distribution for different working fluids which are water, ethanol and acetone for different heating power of 100.14W and 275.6W. From the figure, the temperature fluctuation of the evaporator increases for different working fluids at different heating power. Moreover, the $T_{eva,\ avg}$ also increases when the working fluids change from water to ethanol and acetone. This is attributed to the formation of bubbles during boiling process, increasing the temperature fluctuation in the evaporator. Figure 6(a-b) show the comparison of temperature distribution for inclination angles fluids at 90°, 45° and 20° for different heating input of 100.14W and 275.6W. From the figure, it is obvious that the $T_{eva,\ avg}$ of the TPCT at the inclination angle of 90° is lower than the inclination angles of 45° and 20°. As the inclination angle increases in the range of 20°–90° range, the $T_{eva,\ avg}$ decreases. In the case of the inclination angle of 20°, the formation of bubbles occur on the upper wall of the evaporator, resulting in large wall temperature fluctuations of the evaporator, thus increasing the average temperature. Figure 7(a-b) show the relationship between thermal resistance and heating power for different working fluids and inclination angle. From the figure, the thermal resistances decrease when the working fluids change from acetone to ethanol and water from 0.68 °C/W to 0.56 °C/W and 0.56 °C/W to 0.18 °C/W for the evaporators when the heating power is increased from 100.14W to 275.6W. A notable reduction of thermal resistance can be observed when the inclination angle is increased from 20° to 90°. From this analysis, it can be concluded that the best working fluid and inclination angle with regard to the thermal resistance were found to be water and 90°, respectively, depending on the heating power.

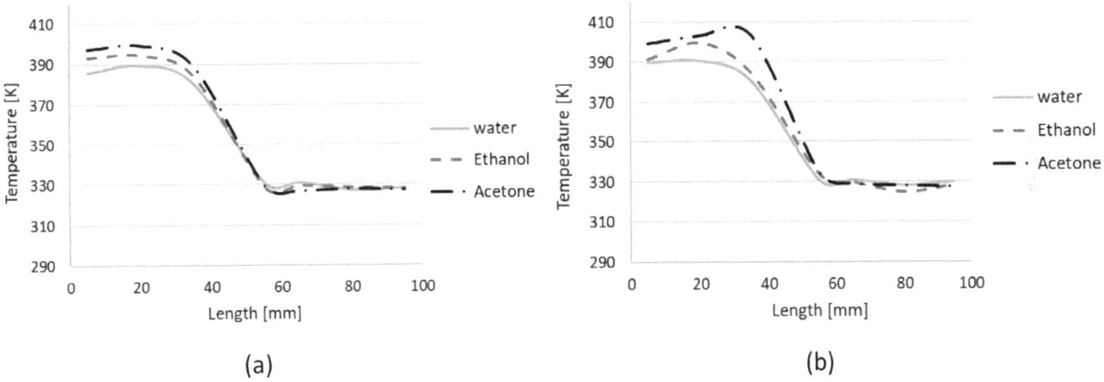

Figure 5. Comparison of temperature distribution for different working fluids for heating power of (a) 100.14W and (b) 275.6W

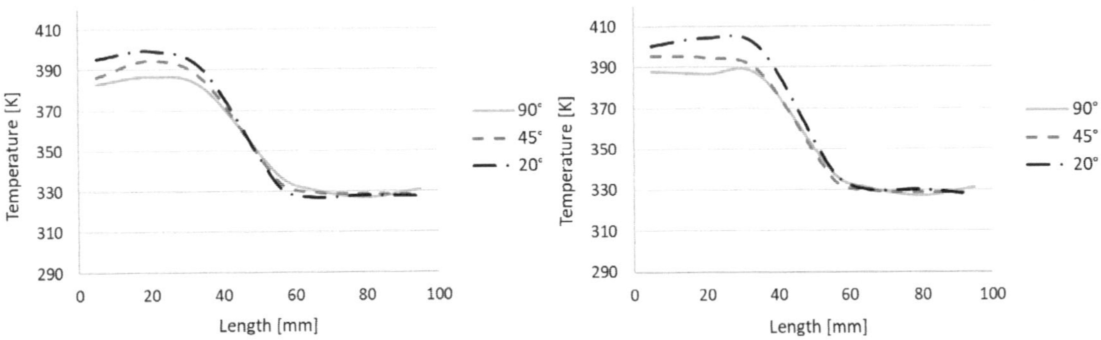

Figure 6. Comparison of temperature distribution for different inclination angles for heating power of (a) 100.14W and (b) 275.6W.

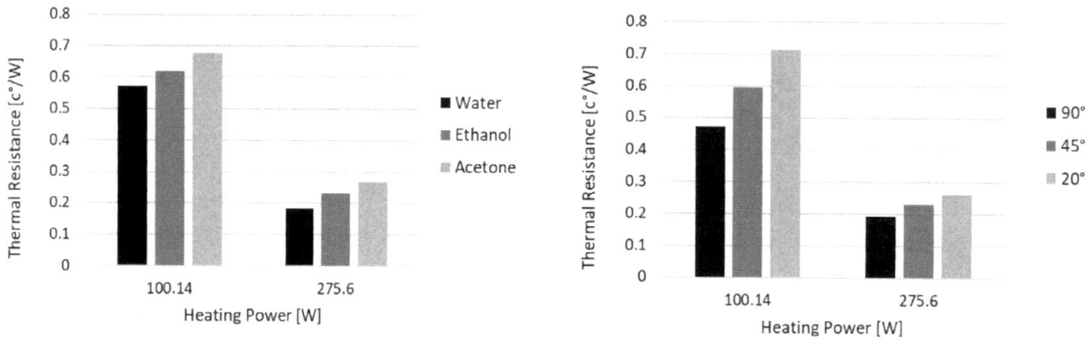

Figure 7. Relationship between thermal resistance and heating power for different (a) working fluids and (b) inclination angle.

4. Conclusion

This paper investigates the thermal performance of a two-phase closed thermosyphon (TPCT) with different working fluids and inclination angles. A two-dimensional CFD model of copper TPCT is simulated with different working fluids (water, ethanol and acetone) at different inclination angles (90°, 45° and 20°) with different heating powers of 100.14W and 275.6W. The effects of the working fluids and inclination angles on the dynamic behaviour of thermal characteristics of TPCT are investigated using ANSYS Fluent. The simulation results showed that ANSYS Fluent with the VOF model is able to model the complex phenomena inside the TPCT, including the boiling of fluid pool in the evaporator and the formation of condensed liquid film in the condenser. The average temperature of the evaporator and condenser have been compared with the published data, showing that there is a good agreement between present CFD results and published data. From the simulation results, it can be concluded that the thermal performance of TPCT is influenced by its type of working fluid and inclination angle. The best working fluid and inclination angle with regard to the thermal resistance were found to be water and 90°, respectively. Moreover, it is found that increasing the heating power inputs from 100.14 to 275.6 W has improved the thermal performance of the TPCT.

5. References

[1] Surip W, Putra N, Antariksawan AR. Design of passive residual heat removal systems and application of two-phase thermosyphons: A review. Progress in Nuclear Energy 2022;154:104473. https://doi.org/10.1016/j.pnucene.2022.104473.

[2] Robinson AJ, Smith K, Hughes T, Filippeschi S. Heat and mass transfer for a small diameter thermosyphon with low fill ratio. International Journal of Thermofluids 2020;1–2:100010. https://doi.org/10.1016/j.ijft.2019.100010.

[3] Jafari D, Franco A, Filippeschi S, di Marco P. Two-phase closed thermosyphons: A review of studies and solar applications. Renewable and Sustainable Energy Reviews 2016;53:575–93. https://doi.org/10.1016/j.rser.2015.09.002.

[4] Kim M, Moon JH. Numerical investigation and modeling of thermal resistance and effective thermal conductivity for two-phase thermosyphon. Case Studies in Thermal Engineering 2021;27:101358. https://doi.org/10.1016/j.csite.2021.101358.

[5] Xu Z, Zhang Y, Li B, Huang J. Modeling the phase change process for a two-phase closed thermosyphon by considering transient mass transfer time relaxation parameter. Int J Heat Mass Transf 2016;101:614–9. https://doi.org/10.1016/j.ijheatmasstransfer.2016.05.075.

[6] Fadhl B, Wrobel LC, Jouhara H. Numerical Modelling of the Temperature Distribution in a Two-Phase Closed Thermosyphon. 2013.

[7] Anto LP, Varghese J. Three-dimensional steady state numerical analysis of inclined two phase closed thermosyphon for solar applications. Case Studies in Thermal Engineering 2022;30:101805. https://doi.org/10.1016/j.csite.2022.101805.

[8] Varun K, Arunachala UC, Vijayan PK. Experimental demonstration of the performance of the novel thermosyphon heat transport device and comparison with CFD predictions. International Journal of Thermal Sciences 2022;176:107503. https://doi.org/10.1016/j.ijthermalsci.2022.107503.

[9] Höhne T. CFD simulation of a heat pipe using the homogeneous model. International Journal of Thermofluids 2022;15:100163. https://doi.org/10.1016/j.ijft.2022.100163.

[10] Kim Y, Shin DH, Kim JS, You SM, Lee J. Boiling and condensation heat transfer of inclined two-phase closed thermosyphon with various filling ratios. Appl Therm Eng 2018;145:328–42. https://doi.org/10.1016/j.applthermaleng.2018.09.037.

Acknowledgments
The authors would like to acknowledge the Ministry of Higher Education of Malaysia for the financial support under the Fundamental Research Grant Scheme (FRGS/1/2019/TK03/UKM/02/6).

Indoor Performance Analysis of a Novel Double-Pass photovoltaic/thermal (PV/T) Asymmetric Compound Parabolic Concentrator (ACPC) Solar Collector

Wan Nur Adilah Wan Roshdan[1], Hasila Jarimi[1*], Adnan Ibrahim[1], Kamaruzzaman Sopian[1], Ali H.A. Al-Waeli[2]

[1]Solar Energy Research Institute (SERI), Universiti Kebangsaan Malaysia, 43600 Bangi, Selangor, MALAYSIA.

[2]Department of Engineering, American University of Iraq - Sulaimani (AUIS), 46001 Kirkuk Rd, Sulaymaniyah, IRAQ

*Email : hasila.jarimi@ukm.edu.my

Abstract. A photovoltaic/thermal (PV/T) solar collector combines solar photovoltaic (PV) modules and solar thermal that operates simultaneously to generate electricity and thermal energy. Design parameters, operating conditions and environmental factors all have a significant impact on the performance of a PV/T solar collector. This research investigates the influence of the variation in the air mass flow rate to the performance of a novel double-pass photovoltaic/thermal (PV/T) asymmetric compound parabolic concentrator (PV/T-ACPC) solar collector. To investigate performance of the solar collector, it was tested indoors using a solar simulator at average solar radiation of 800 W/m² with the variation in the air mass flow rate ranging from 0.0074 kg/s to 0.0900 kg/s. From the analysis, we found that as the air mass flow rate increases, the thermal efficiency and electrical efficiency increases from 37.15 % to 60.51 %, and 2.51 % to 3.29 % respectively. Meanwhile while the difference in the air output temperature and PV panel temperature were found to decrease from 19.64 °C to 2.63 °C, and 66.83 °C to 41.86 °C respectively. We also found that, as the mass flow rate continues to increase, it will reach its 'optimum point' and approaching a plateau at mass flow rate of 0.0262 kg/s. The finding is crucial since the collector needs to be operated at its optimum flow rate to ensure optimum efficiencies at the optimum temperature rise of the useful air supply.

Keywords: Photovoltaic/thermal (PV/T) collector, Asymmetric Compound Parabolic Concentrator (ACPC), Air mass flow rate, Thermal efficiency, Electrical efficiency

1. Introduction

Photovoltaic/thermal (PV/T) hybrid collectors combines the mechanism of solar photovoltaic technologies and solar thermal absorbers. It has the potential as a power generation technology that are able to convert solar radiation into usable electrical and thermal energy onto the same collecting surface [1]. In addition, the thermal component of the collector can maximize the PV cell efficiency by heat removal system. The design configuration, material selection and working fluid can affect the overall performance of the PV/T collector.

There are many different working fluids that is suitable to be integrated into a PV/T collector. Researchers have investigated different working fluids such as air, water, bi-fluid [2] and other advanced materials such as PCM and nanofluid [3], [4]. In terms of preparation, lower cost and easier maintenance, air is the most cost-effective working fluid. Nevertheless, air solar collector has a low heat transfer rate due to its low thermal conductivity properties and higher volume of air per unit area is required [5]. Hence, in order to address this issue, a double pass configuration can enhance the convective heat transfer that occurs within the collector due to an increase heat transfer area [6], [7].

Concentrating the collector by intensifying the solar radiation incidents from large area to smaller absorber area offers to address the issues of the high cost of the semiconductor material which is the main component of PV panels [8]. The compound parabolic concentrator (CPC) solar collectors have emerged as a viable option for several applications in the field of solar energy due to its capacity to collect both direct and diffuse solar radiation onto a collector and is also suitable for stationary installation [9]. However, due to different geographical location, the solar incidence angle is not symmetrical and constant throughout the year. To encounter this matter, asymmetric configuration of the parabola with different acceptance angle can maximize the potential to receive the solar radiation without having to adjust the collector at a tilt or employing high cost of solar tracking [10].

According to the previous references published by the above-mentioned researchers, the influence on PV/T solar collector may give diverse characteristics depending on numerous aspects of the design parameters. An asymmetric CPC PV/T double pass air solar collector is investigated in the present study. The effect of air mass flow rate to the performance of the collector is further discussed in this paper.

2. Design of Double-Pass PV/T ACPC Air Solar Collector

2.1. Optimum Acceptance Angle of the ACPC Geometry

The following researchers, Wan Roshdan et al [10] have investigated the optimum pair of acceptance (θ_{CI} and θ_{CII}) angles for the Double-Pass photovoltaic/thermal (PV/T) Asymmetric Compound Parabolic Concentrator (ACPC) Solar Collector when installed at different locations namely; Kuala Lumpur, Malaysia, Abu Dhabi, UAE, London, UK and Bergen, Norway, representing low, middle and high latitude countries respectively. It is found that the optimum pair of acceptance angle are 5° and 65° for low, mid and high latitude countries and 5° and 55° for another area of high latitude countries. Figure 1 shows the total yearly absorbed solar irradiation for different pairs of half acceptance angle.

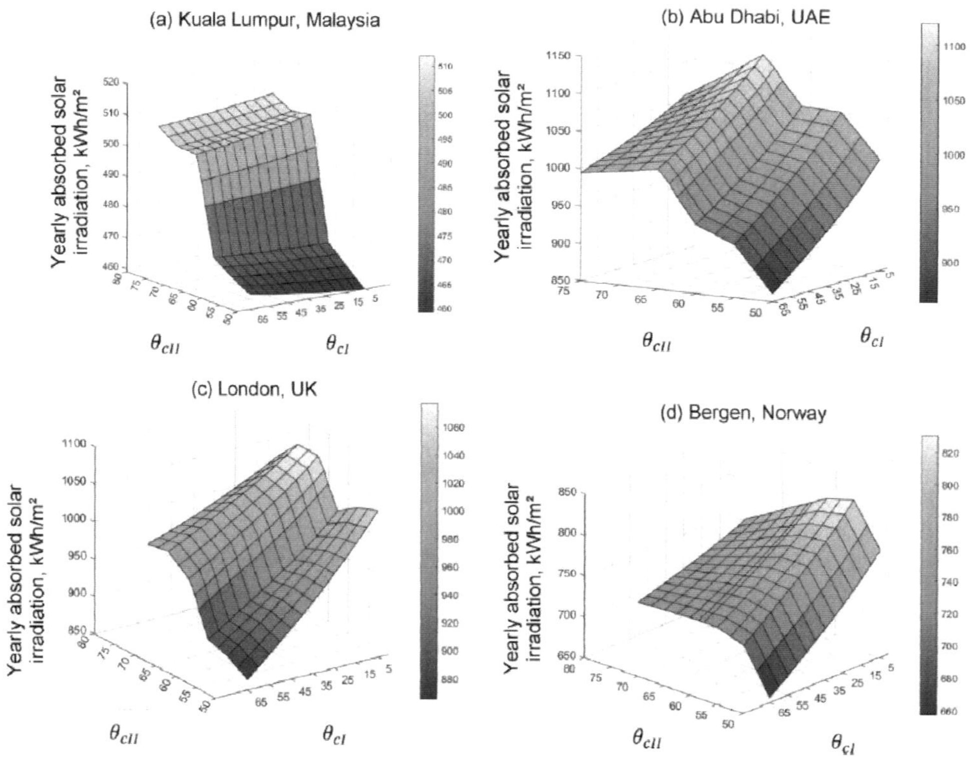

Figure 1. Total yearly absorbed solar irradiation for different pairs of half acceptance angle for (a) Kuala Lumpur, Malaysia (b) Abu Dhabi, UAE (c) London, UK and (d) Bergen, Norway [10].

2.2. Design of the Double-Pass PV/T ACPC solar collector

Three PV panels of which each with 32 monocrystalline PV cells at 0.12 m x 0.95 m was used in this study. Hence, the total area of the PV panel is 0.342 m^2. The specifications of the PV panel are also presented in Table 1.

Table 1. Specification of each PV panel

Parameters	**Specification**
Pmax	17W
Vmp	16V
Imp	1.06A
Voc	19.2V
Isc	1.28A

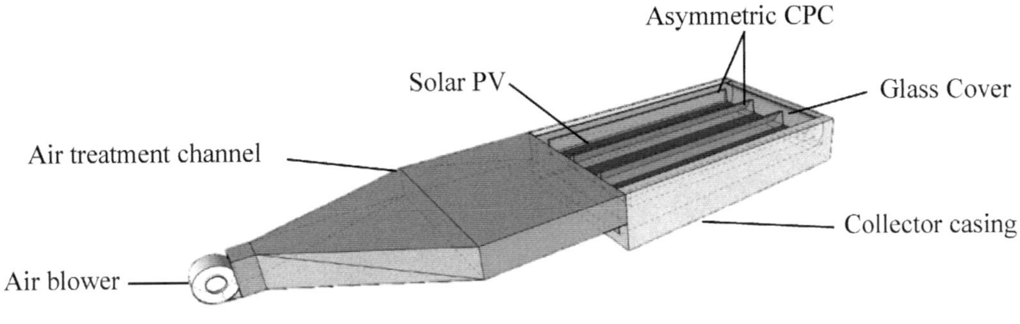

Figure 2. The design of the collector

Figure 2 shows the collector design. The collector has a glass cover on top of the collector. Under the glass cover, there are three solar PV panels that are attached together with a pair of ACPC at each side of the panel in order for the solar radiation to undergo reflection and will be concentrated onto the surface of the panel. The air will be forced to enter the collector, passing through the top of the absorber, flowing back to the bottom of the PV panels and exit the collector. The temperature of the PV cell that rises due to the concentration that occurs, will reduced due to the convection heat transfer mechanism that occurs throughout the collector. The heated air is useful for thermal energy for heating application.

3. Materials and Method

3.1 Experimental procedures

The PV/T-ACPC solar collector was placed and tested indoor under controlled environment in laboratory of Solar Energy Research Institute (SERI), Universiti Kebangsaan Malaysia. The solar simulator of 8 x 50 W halogen lamps was placed perpendicular to the collector at height of 0.6 m and the intensity was set to average solar radiation of 800 W/m². Three 17 W solar panels were connected in parallel and the output of the solar PV connectors were connected to an I-V Tracer to obtain electrical data of the collector. Thermocouples of K-Type were placed at several locations within the collector to measure the temperature of the PV cells, inlet and outlet of fluid temperature, ambient temperature and the absorber of the collector. A data acquisition is used to record the data from the thermocouple of every 60 seconds. Forced convection of air was introduced to an air treatment channel before entering the inlet part of the collector. The mass flow rate of the air was varied from 0.0074 kg/s to 0.0900 kg/s and it was regulated by controlling the frequency at the blower. The flowchart methodology for this experiment is presented in Figure 3. Figure 4 shows the diagram of the indoor experimental setup.

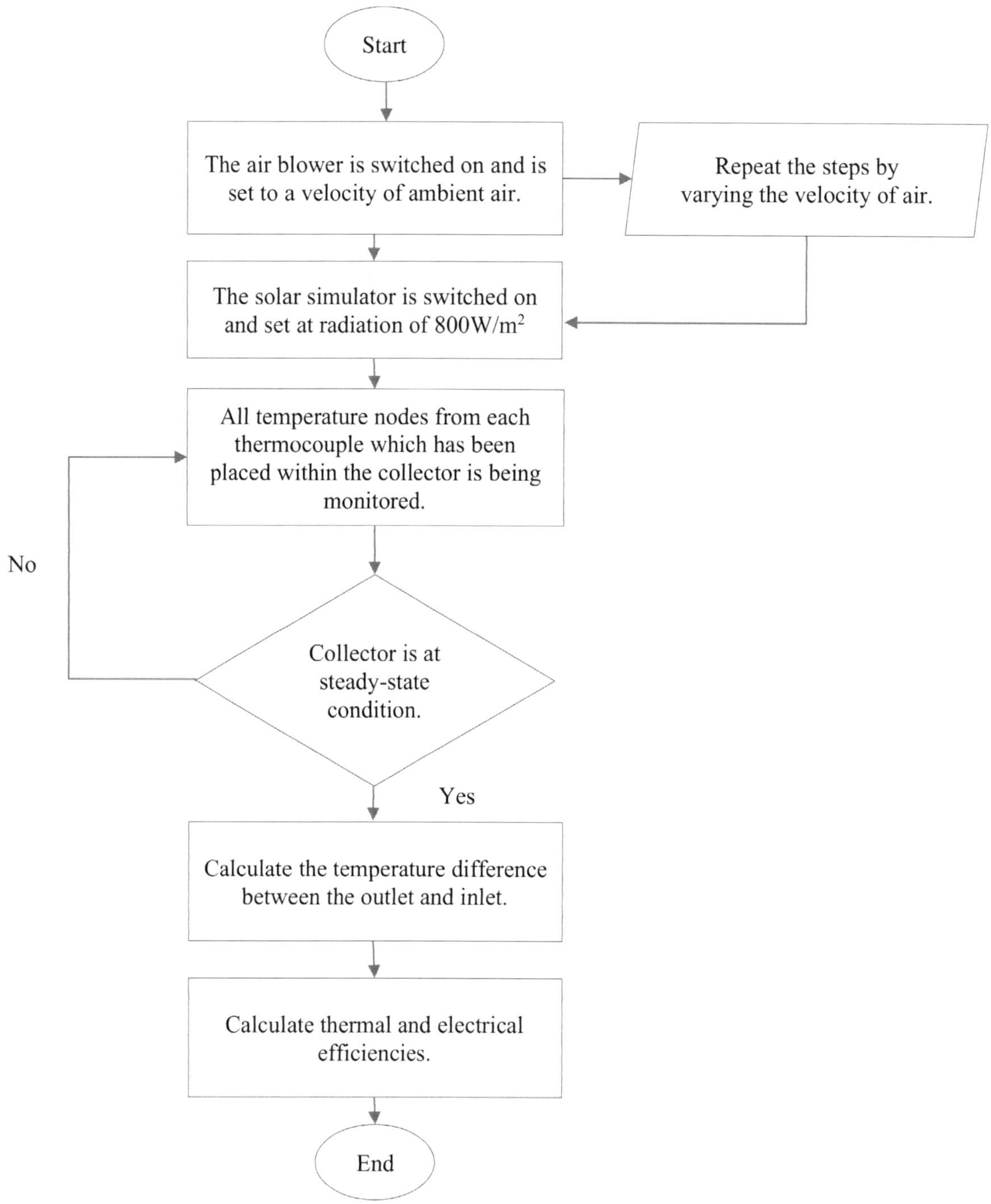

Figure 3. The methodology flowchart to calculate the thermal and electrical performance

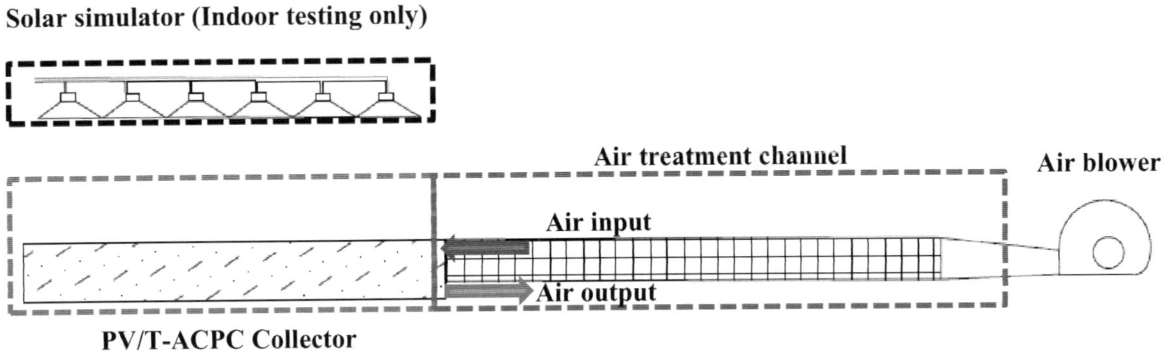

Figure 4. Diagram of indoor experimental setup

3.2 Performance of PV/T-ACPC solar collector

In this study, the performance of the thermal and electrical were measured based on the working fluid and surface temperature of the PV collector. Thermal efficiencies of the collector were calculated using the equations as below:

$$\eta_{th} = \frac{\dot{m}\, C_p(T_{fo} - T_{fi})}{A\, G} \tag{1}$$

Where the thermal efficiencies are represented by, $\eta_{th}(\%)$, mass flow rate, $\dot{m}$ (kg/s), specific heat capacity of air C_p (J/kg. °C), inlet temperature, T_{fi} (°C), outlet temperature, T_{fo} (°C), area of the collector, A (m^2) and intensity of the radiation, G (W/m^2). Meanwhile, for the electrical efficiency of the collector can be presented using the following equation:

$$\eta_{ele} = \frac{P_{max}}{A\, G} \tag{2}$$

Where the parameters in equation (2) is are $\eta_{ele}(\%)$, maximum power, P_{max} (W), area of the PV surface, A (m^2) and intensity of the radiation, G (W/m^2).

4. Results and Discussion

4.1 PV/T-ACPC Solar Collector Thermal Performance

Figure 5 shows the influence of air mass flow rate to the change in the temperature rise and the thermal efficiency of the collector. The increase in the air mass flow rate implying that volume of air in removing the heat from the solar collector increases per second. It also indicates an increase in the overall convection heat transfer from the inner surfaces of the PV panel to the working fluid and hence the thermal efficiency of the solar collector was found to increase from 37.15% to 60.51%. In addition, as the air mass flow rate increases, the trend in the recorded air temperature rise was found to decrease from 19.64 °C to 2.63 °C. This is attributed by the fact that air spends a shorter time in the air channel as its mass flow rate increases.

It is also crucial to note that, the solar collector could be operated at its maximum flow rate which leads to high efficiency but a small increase in fluid temperature and high requirement of pumping power. This implies that, the collector needs to be operated at its optimum flow rate to ensure optimum efficiencies at the optimum temperature rise and hence, determining the optimum mass flow rate of the working fluid is crucial. In this study, the optimum flow rate can be determined from Figure 5. As presented in Figure 5, the increase in thermal efficiency and the decrease in temperature are observed until it reaches an optimum point whereby further increase in mass flow rate leads to no further changes to these two parameters. From the analysis, the optimum mass flow rate was found at 0.0262 kg/s with the measured thermal efficiency and air temperature difference of 56.73 % and 8.47 °C respectively.

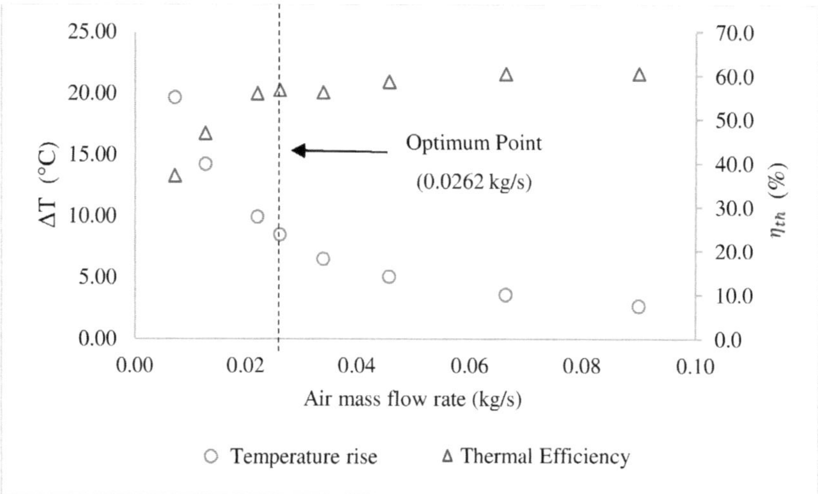

Figure 5. The Influence of Air Mass Flow Rate to the Thermal Efficiency (η_{th}) of the Solar Collector [G= 800 W/m^2]

4.2 PV/T-ACPC Solar Collector Electrical Performance

Figure 6 shows the influence of air mass flow rate to the electrical performance of the collector. In the steady-state condition, increasing air mass flow rate improved the cooling of PV cells; hence, the mean temperature of PV cells, T_{PV} decreases. Consequently, due to the inverse relationship of the electrical efficiency with the temperature of the PV cells, the electrical performance of the collector improved. From the experiment, it is observed that the temperature of PV cells dropped from 66.83 °C to 41.86 °C. The decrease in the PV cell temperature increases the electrical efficiency from 2.51 % to 3.29 %. From the experiment, increasing the air mass flow rate has demonstrated that the electrical efficiency increases aligned with the thermal efficiency.

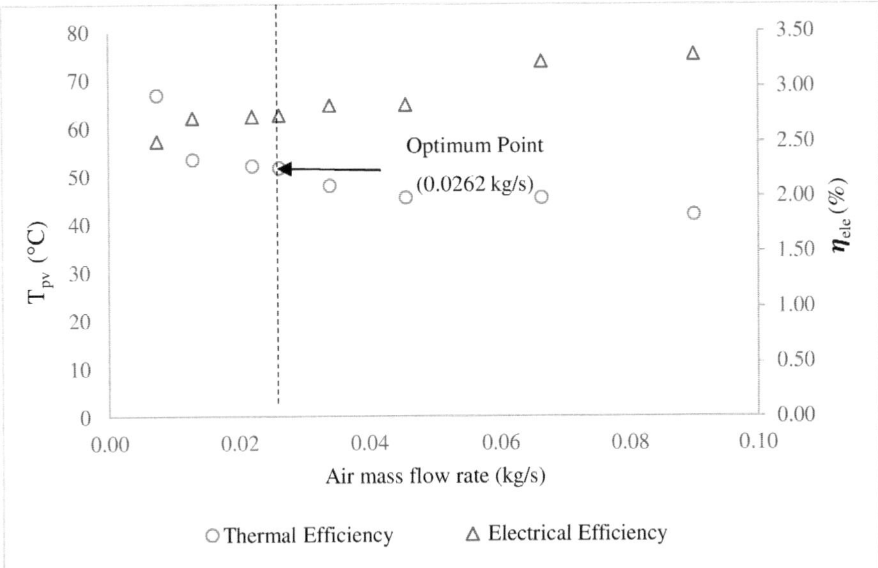

Figure 6. The Influence of Air Mass Flow Rate to the Electrical Efficiency (η_{ele}) of the Solar Collector [G= 800 W/m^2]

5. Conclusion

Based on the present study, the following conclusions can be drawn. The variation in the air mass flow rate does influence the performance of a novel double-pass photovoltaic/thermal (PV/T) asymmetric compound parabolic concentrator (PV/T-ACPC) solar collector. As the air mass flow rate increases, the temperature difference in the air output temperature decreases from 19.64 °C to 2.63 °C and the PV panel temperature decreases from 66.83 °C to 41.86 °C. Due to this decrease in temperature difference, the thermal efficiency and electrical efficiency increases from 37.15% to 60.51%, and 2.51% to 3.29% respectively. To ensure optimum efficiencies, the collector must be operated at optimum temperature and flow rate. As the mass flow rate continues to increase, it will reach its 'optimum point' and approaching a plateau. Therefore, in this experiment, the 'optimum point' was found to be at mass flow rate of 0.0262 kg/s.

References

[1] A. H. A. Al-Waeli, H. A. Kazem, M. T. Chaichan, and K. Sopian, *Photovoltaic/thermal (PV/T) systems: Principles, design, and applications*. 2019.

[2] A. Sayigh, "Mediterranean green buildings & renewable energy: Selected papers from the world renewable energy network's med green forum," *Mediterr. Green Build. Renew. Energy Sel. Pap. from World Renew. Energy Network's Med Green Forum*, pp. 1–963, 2017

[3] J. Akhter, S. I. Gilani, H. H. Al-Kayiem, M. Ali, and F. Masood, "Experimental evaluation of thermophysical properties of oil-based titania nanofluids for medium temperature solar collectors," *Materwiss. Werksttech.*, vol. 51, no. 6, pp. 792–802, 2020

[4] J. Akhter, S. I. Gilani, H. H. Al-kayiem, M. Ali, and F. Masood, "Characterization and stability analysis of oil-based copper oxide nanofluids for medium temperature solar collectors," *Materwiss. Werksttech.*, vol. 50, no. 3, pp. 311–319, 2019

[5] L. Sahota and G. N. Tiwari, "Review on series connected photovoltaic thermal (PVT) systems: Analytical and experimental studies," *Sol. Energy*, vol. 150, pp. 96–127, 2017

[6] A. F. Ismail, A. S. A. Hamid, A. Ibrahim, H. Jarimi, and K. Sopian, "Performance Analysis of a Double Pass Solar Air Thermal Collector with Porous Media Using Lava Rock," *Energies*, vol.

15, no. 3, 2022

[7] Amrizal, Y. P. Hidayat, and M. Irsyad, "Performance comparison of single and double pass PV/T solar collectors integrated with rectangular plate fin absorber," *IOP Conf. Ser. Mater. Sci. Eng.*, vol. 1173, no. 1, p. 012023, 2021

[8] F. Masood *et al.*, "A Review of Recent Developments and Applications of Compound Parabolic Concentrator-Based Hybrid Solar Photovoltaic / Thermal Collectors," 2022.

[9] R. V. Parupudi, H. Singh, and M. Kolokotroni, "Low Concentrating Photovoltaics (LCPV) for buildings and their performance analyses," *Appl. Energy*, vol. 279, no. September 2020, p. 115839, 2020

[10] W. N. A. Wan Roshdan, H. Jarimi, A. H. A. Al-Waeli, O. Ramadan, and K. Sopian, "Performance enhancement of double pass photovoltaic/thermal solar collector using asymmetric compound parabolic concentrator (PV/T-ACPC) for façade application in different climates," *Case Stud. Therm. Eng.*, vol. 34, no. February, p. 101998, 2022

Acknowledgments

The authors would like to thank Universiti Kebangsaan Malaysia for Young Lecturers Incentive Grant (GGPM) with grant number GGPM-2020-032.

IOP Conf. Series: Materials Science and Engineering 1278 (2023) 012010 doi:10.1088/1757-899X/1278/1/012010

Distribution cycle of wind speed: A case study in the Southern Part of Malaysia

Daniel Derome[1,3], Halim Razali[1], Ahmad Fazlizan[1], Alias Jedi[2]

[1]Solar Energy Research Institute, Universiti Kebangsaan Malaysia, 43600 Bangi Selangor Malaysia
[2]Faculty of Engineering and Built Environment, Universiti Kebangsaan Malaysia, 43600 Bangi Selangor Malaysia
[3]Department of Mathematics, Science & Computer, Politeknik Sultan Idris Shah, 45100 Sabak Bernam Selangor Malaysia

Abstract: Wind energy has historically helped many people and has been utilized for farming and seafaring. Today, modern usage is more comprehensive such as generating electricity. However, its usage in Malaysia remains at an all-time low. Numerous factors contribute to this result. Consequently, a full assessment of wind energy in a particular region is deemed necessary. This study investigates wind speed distribution in Mersing. It also illustrates the period during which the regional distribution statistics changed. The researcher used Autofit 5.6 and the Maximum Likelihood Method to figure out the parameter value of each distribution for wind speed data from 2012 to 2016. This study utilized three distinct methods of Goodness of fit (GOF) to estimate the optimal distribution ranking, such as Chi-Square, Anderson-Darling, and Kolmogorov-Smirnov. The results reveal that the Burr, Gamma, and Lognormal distributions correlate to the wind speed data in Mersing and the Burr distribution is the most appropriate. In addition, the findings indicate that distribution changes occur at a minimum rate of 8.89%. In conclusion, these findings reveal that there has been no significant change over the five years. This information enables stakeholders to determine whether they will earn a profit owing to changes in the distribution of a particular area, reducing risk.

1. Introduction

Late in the 1990s, the government of Malaysia implemented renewable energy. Since the 8th Malaysian Plan, the government has promoted renewable energy in general and wind energy in particular [1]. This optimism may be attributable to the widespread use of wind energy in countries with terrains comparable to Malaysia, such as Thailand [2] and India [3]. Wind energy is an unlimited source of renewable energy [4], and it was among the most advanced technologies accessible at the time [5]–[8]. The initial wind energy project in Malaysia was successful, and Pulau Layang-Layang in Sabah was the first location chosen by the government [9]. In general, however, Malaysia's wind energy production lags behind other renewable energy sources, such as solar. As a result, Malaysia's wind energy production is moving at a snail's pace compared to other renewable energy sources.

Malaysia produces almost no wind energy [10]. A lack of public understanding of the significance of wind energy generation is one of the problems restricting the growth of wind energy in Malaysia. This condition does not attract interested parties, such as private parties, outsiders, or potential investors [11]. This deficiency is attributable to the fact that these stakeholders are sceptical and have some fear regarding wind power. As a result, people and private businesses are afraid to take risks in Malaysia, where wind energy research is incomplete. This risk is unresolved due to the difficulty in selecting the best appropriate place, which remains a conundrum and a mystery. Before establishing any wind energy system, it is essential to do extensive research and communicate wind data. The principal application of the distribution is forecasting wind speed. Next, the distribution must be studied because these alterations correlate highly with investment return (ROI). In addition, the potential stakeholders spent a substantial amount of the funds. Consequently, the period of distribution shift is crucial to consider.

In the study of wind speed data, various distributions have been utilized. Worldwide, the Weibull distribution is linked with wind speed research [12]. Therefore, it must be identified for all utilized data. Numerous studies in Thailand [13], Yemen [14], Bangladesh [5], Senegal [7] and Malaysia [15] determined that the Weibull distribution best matched their data. Thus, researchers typically employed the Weibull distribution as a standard [16], [17]. However, not all regions exhibit the Weibull distribution form [18], [19]. For example, a Malaysian study [20] revealed that the Burr distribution provides the best match. Unfortunately, [21] reveals that the Gamma distribution is appropriate for Malaysian wind speed data. This conclusion is supported by [22], the best match for Malaysia's twenty stations. However, in 2016, [23] determined that Malaysia's Weibull and Gamma distributions were optimal. This result is attributable to the Weibull distribution being less favourable than other distributions when a region's wind speed data has a large proportion of zero values. Additionally, the Weibull distribution is not the optimal fit for data with several modes. [19], [24] also concurred on these two aspects.

Wind velocity is the critical factor influencing data distribution in a specific region [25], [26]. It influences data distribution since wind speed is one of the essential elements in wind energy research. Consequently, the dispersion will vary according to the wind speed in a particular place. When will the distribution begin to change? Unfortunately, as far as the author is aware, there is virtually no research examining the duration of the dispersion in Malaysia. As reported by [27], [28], this data distribution is essential because it is the initial step in performing additional research.

Before establishing a wind energy system, it is essential to undertake a study on the distribution of wind data [7], [16], [24], [29]. Consequently, a thorough inquiry should be made. The study aims to examine the wind speed distribution in Mersing and determine the duration of change in the area's data distribution. This study is significant because it will evaluate whether the change period is shorter or longer than the return on investment of a wind turbine if one is erected [30].

2. Method

Mersing (Figure 1) is a location commonly recommended by past researchers. It was selected as the study site based on the findings of prior [23], [31], [32] studies. In addition, the data is trustworthy and of excellent quality.

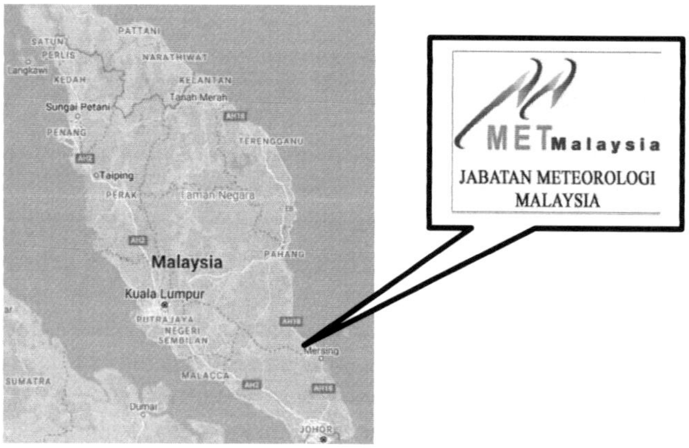

Figure 1. Location of Mersing (Image from Google Maps)

Calculating the daily average wind speed yields wind velocity information. Consequently, the sample size is comparable. There were, however, years in which the number of samples collected declined marginally. This decrease in samples results from technical issues at the weather station. According to an examination of the relevant literature, [6] more than seventy-five per cent of the data can be used for the research with some adjustments.

Then, using the software Autofit 5.6, the data were analyzed. The type of distribution is determined using the literature review results. Weibull, Gamma, Normal, Lognormal, Binomial, Poisson, Rayleigh, Burr, Gauss, and Exponential are among the more than ten well-known distributions linked with wind speed distribution studies. This study uses the Maximum Likelihood Method (MLM) to calculate parameter values for each wind speed data distribution.

In addition, the results were analyzed using the Goodness of Fit technique (GOF). GOF seeks to identify the ranking of the best distributions. Moreover, the previous researcher suggests employing numerous GOFs for position ranking [33].
The formula for Chi-Square as in (1),

$$\chi^2 = \frac{(O_i - E_i)^2}{E_i} \quad (1)$$

where O_i is the observed data and E_i is the expected data [33].

While the formula for Anderson-Darling (AD) is as in (2),

$$AD = -n - \frac{1}{n}\sum_{i=1}^{n}(2i-1)[\ln F(X_i) + \ln(1 - F(X_{n-i+1}))] \quad (2)$$

Where n =the sample size, $F(v)$ is the cumulative probability distribution of the specific distribution and i= the ith sample, calculated when the data is sorted in ascending order [34].

The formula for Kolmogorov-Smirnov is as in (3),

$$KOL = Max|F(v) - F_n(v)| \quad (3)$$

where v is the identifies the set of the velocity to be considered, $F(v)$ is the cumulative probability distribution for specific distribution and $F_n(v)$ is the experimental histogram [35].

3. Results and Discussion

The descriptive statistic of the measurement data was created, as shown in Table 1. Based on the average mean value, the annual wind speeds in Mersing are comparable to and moderate in their intensity [36]. According to the International Electrotechnical Commission (IEC), low wind speeds are between 0 and 7.5 meters per second. [37].

Skewness offers a first look at wind speed data by swaying it to the right or left, providing a preliminary perspective on the data. In the meantime, the kurtosis gives information about the peaks of the graphs. For instance, all of the values for skewness and kurtosis in Table 1 are depicted as having a positive sign. The fact that this skewness number is positive indicates that there may be significant positive outliers within the data on wind speed. Similarly, if the kurtosis value is positive, it indicates a prominent peak in the graph. This value demonstrates how the data regarding the wind are condensed into a smaller range. As a result, the year 2012 is the most suitable for ranking. The minimum value variance and the coefficient of variation are used to determine this rating (COV).

Table 1. Descriptive statistics of the measured data.

	2012	**2013**	**2014**	**2015**	**2016**
n	366	365	365	365	366
Mean	2.6164	2.7088	2.8452	2.9932	2.8306
Variance	0.6212	0.7753	0.8809	0.8297	1.0332
COV	0.3012	0.3251	0.3299	0.3043	0.3591
Skewness	1.9074	1.4205	1.4054	1.3167	1.9180

Kurtosis	4.3234	1.6993	1.3926	1.5510	3.9537
Min	1.5	1.3	1.5	1.6	1.5
Max	6.1	5.8	5.9	6.2	7.6

The histogram and probability density function (pdf) of the data from 2012 to 2016 are presented in Figure 2. While the probability density function characterizes an estimate of the data, the histogram displays the actual data that was collected (model). Only the results show a correlation between the top three distributions and the wind speed data collected in Mersing throughout the study. The Burr, Gamma, and Lognormal distributions are examples of optimal fits that can be found here. As a consequence of this, each distribution that corresponds to Mersing has three parameters. The information presented in Table 2 and summarized in Table 3 pertains to Figure 2.

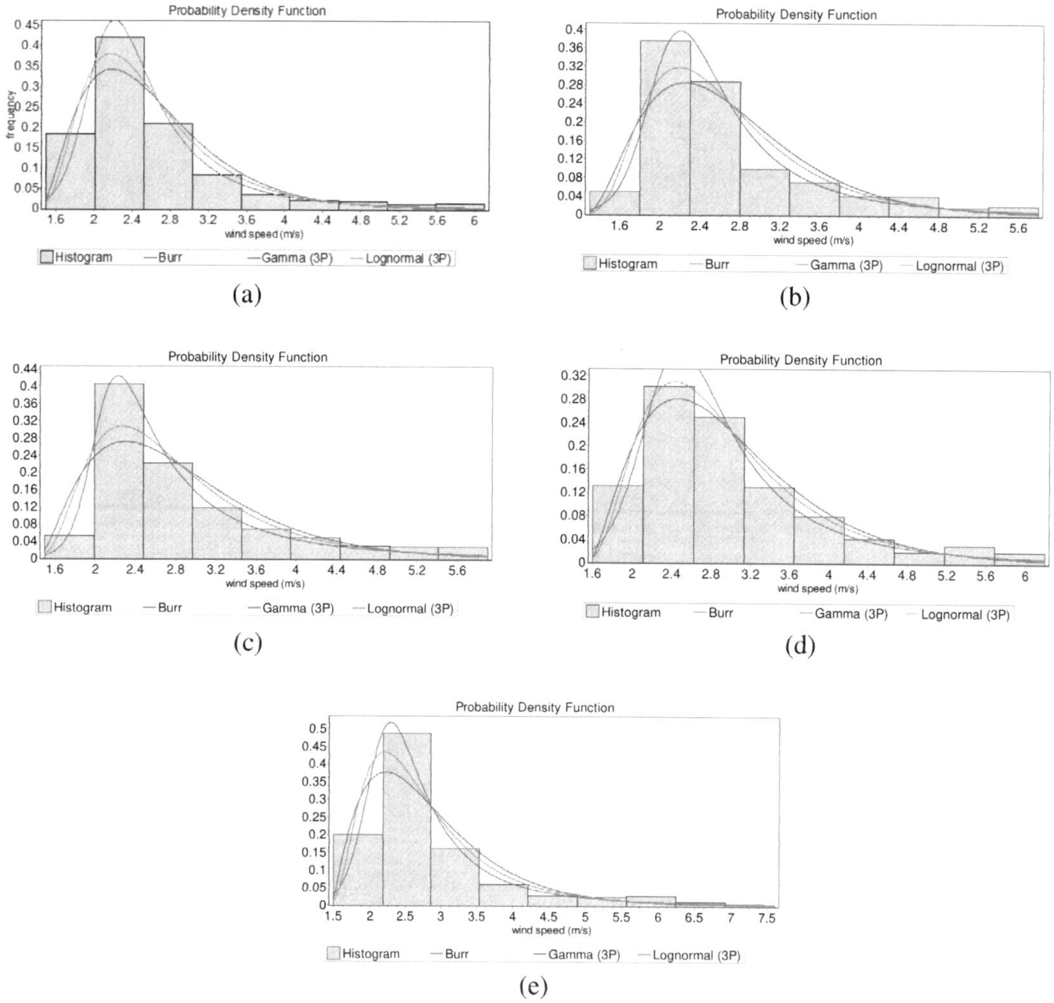

Figure 2. Histogram and probability density function a) 2012 b) 2013 c) 2014 d) 2015 e) 2016

Table 2 presents the findings obtained from the Chi-Square and Anderson-Darling tests, which were conducted to investigate the variation in wind speed from 2012 to 2016. The purpose of these tests was to determine the optimal distribution arrangements, and the outcomes of these tests are combined because they produced the same findings in each instance. Only three distributions, specifically the Lognormal, Burr, and Gamma distributions, appear suitable for use with the Mersing model, as indicated by the preliminary research findings.

In 2012, 2013, 2014, and 2016, the Burr distribution provided the most accurate depiction of the Mersing data, followed by the Lognormal and Gamma distributions. However, there was substantial variation in the distribution arrangements in the fourth year of data (2015). In contrast, the best fit was shown by the Lognormal distribution, followed by the Burr and Gamma distributions. This mismatch is probably attributable to the fact that 2015 experienced the highest extraordinary average wind speed. In light of the findings of the Chi-Square and Anderson-Darling tests, the first inference that can be drawn is that the data distribution has, for the most part, remained the same throughout the five-year study.

Table 2. Best fit distribution by Chi-Square / Anderson Darling

Years	Distribution		
	1st	2nd	3rd
2012	Burr	Lognormal	Gamma
2013	Burr	Lognormal	Gamma
2014	Burr	Lognormal	Gamma
2015	Lognormal	Burr	Gamma
2016	Burr	Lognormal	Gamma

As determined by Kolmogorov-Smirnov (KS), the optimal distribution is depicted in Table 3, which can be found here. The Mersing suitable distribution is similar to Table 2, as shown here. This further demonstrates that the data best fit by three distributions, specifically the Burr, Gamma, and Lognormal distributions. Nevertheless, the Burr distribution offers the most convincing explanation. This disparity is due to the fact that the formulas for KS and Chi-Square/Anderson-Darling are independent of one another. Therefore, according to the findings of the third test (KS), the data distribution did not change throughout the fifth year of research.

Table 3. Best fit distribution by Kolmogorov-Smirnov

Years	Distribution		
	1st	2nd	3rd
2012	Burr	Lognormal	Gamma
2013	Burr	Lognormal	Gamma
2014	Burr	Lognormal	Gamma
2015	Burr	Lognormal	Gamma
2016	Burr	Lognormal	Gamma

The percentage of change in distribution throughout the investigation is illustrated in Figure 3. The data that is presented in Tables 2 and 3 are represented in this image. The positions of 41 of the 45 frequency distributions in both tables are kept the same, while the positions of the other four frequency distributions are switched. The shift accounted for 8.89% of the total change. The wind speed is responsible for this insignificant shift, which is less than 10 per cent.

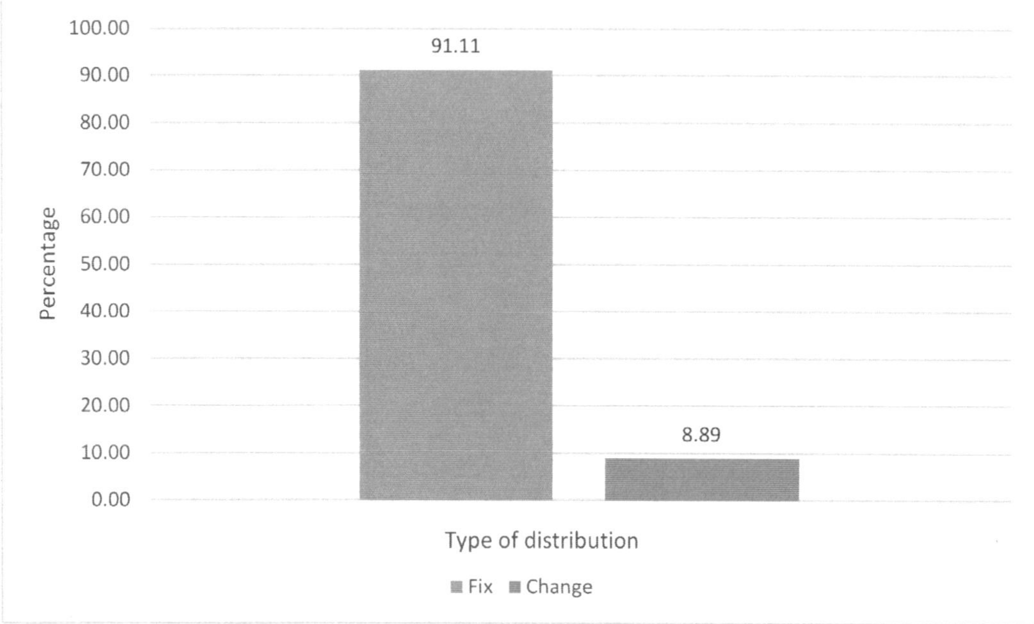

Figure 3. Percentage of distribution change in Mersing

Figure 4 shows that the Mersing distribution is most similar to the Burr distribution, with the Burr distribution accounting for 86.7 per cent of the total distribution. The subsequent distribution that will be discussed is the Lognormal distribution, which accounts for 13.3 per cent of the total. In addition, it is essential to conduct this investigation because it is necessary to know the distribution of distributional changes as well as their duration before acquiring parameters. An accurate parameter estimate for the wind potential of a particular area can be derived from the exact distribution because of how it is distributed. Consequently, determining the wind potential involves going through a series of steps that start with the dispersion calculation.

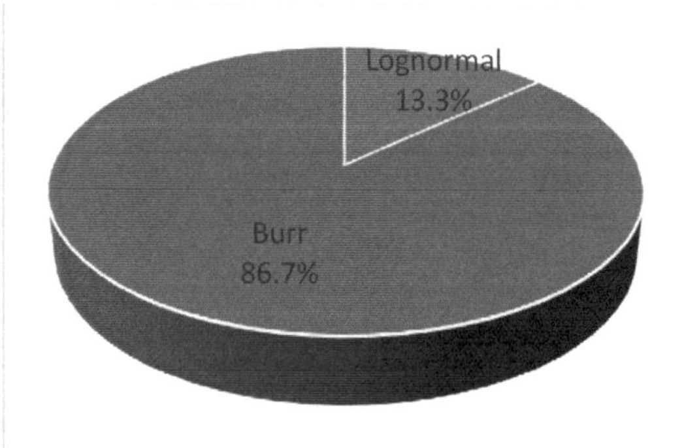

Figure 4. List of the best fit distribution in Mersing

4. Conclusion

This paper presents a study undertaken to explore the Mersing wind speed distribution and establish the duration of change in the area's data distribution. Data is analyzed for five years, from 2012 to 2016. The research results show that the Burr distribution best fits the Mersing data. Aside from that, the location distribution has been reasonably steady over the last five years, with only an 8.89 per cent shift. Although data distribution in the Mersing area is very consistent, several research should be carried out. As a result, one option is to use more extended data periods to determine a distribution cycle at a site. This study has a significant impact because it can be used by stakeholders such as donors, the government, the commercial sector, and other interested parties; this study has a substantial influence. This knowledge will also aid in removing the confusion and rage connected with wind energy failures. Furthermore, the risk is decreased because stakeholders can predict how long it will take for a benefit to accrue due to changes in an area's distribution. Finally, this research should be expanded to realize Malaysia's dream of wind energy generation using all means necessary.

Acknowledgements
The authors would like to thank and appreciate Universiti Kebangsaan Malaysia (UKM) for the research grant allocation under the Grad Challenge Research Project (AP-2015-003) and the Department of Mathematics, Science & Computer, Sultan Idris Shah Polytechnic in all forms of assistance given to prepare this writing.

References

[1] D. Derome, H. Razali, A. Fazlizan, S. Mat, and M. A. Elias, "Wind Speed Distribution : A case study of Mersing , Malaysia," in *Energising Green Buildings, 2nd Malaysia University-Industry Green Building Collaboration Symposium (MU-IGBC 2018), Malaysia Green Building Confederation*, 2018.
[2] C. Smuthkalin, T. Murayama, and S. Nishikizawa, "Evaluation of the wind energy potential of Thailand considering its environmental and social impacts using geographic information systems," *Int. J. Renew. Energy Res.*, vol. 8, no. 1, pp. 575–584, 2018.
[3] D. Sangroya and J. K. Nayak, "Development of wind energy in India," *Int. J. Renew. Energy Res.*, vol. 5, no. 1, pp. 1–13, 2015.
[4] J. Jayapriya, D. Muruganandam, D. Raguraman, B. Senthilkumar, and V. Dhinakaran, "Design and fabrication of multi-rotor horizontal axis wind turbine," *Int. J. Eng. Adv. Technol.*, vol. 8, no. 6, pp. 3500–3504, 2019.
[5] A. K. Azad, M. G. Rasul, M. M. Alam, S. M. A. Uddin, and S. K. Mondal, "Analysis of Wind Energy Conversion System Using Weibull Distribution," *Procedia Eng.*, vol. 90, no. 1, pp. 725–732, 2014.
[6] R. Bothale and Y. Katpatal, "Spatial and Statistical Clustering Based Regionalization of Precipitation and Trend Identification in Pranhita Catchment, India," *Int. J. Innov. Res. Sci. Eng. Technol.*, vol. 3, no. 5, pp. 12557–12567, 2014.
[7] B. Ould Bilal, M. Ndongo, C. M. F. Kebe, V. Sambou, and P. A. Ndiaye, "Feasibility study of wind energy potential for electricity generation in the northwestern coast of Senegal," *Energy Procedia*, vol. 36, pp. 1119–1129, 2013.
[8] L. H. Fang and R. B. A. Rahim, "Design of Savonius model wind turbine for power catchment," *Int. J. Electr. Comput. Eng.*, vol. 12, no. 3, pp. 2285–2299, 2022.
[9] Unknown, "Kincir angin di Pulau Layang-Layang," *Utusan Malaysia*, 1999.
[10] IRENA, "Wind energy," 2017.
[11] H. Cetinay, F. A. Kuipers, and A. N. Guven, "Optimal siting and sizing of wind farms," *Renew. Energy*, vol. 101, pp. 51–58, 2017.
[12] D. Indhumathy, C. V. Seshaiah, and B. Vennila, "Wind speed using improved mixture Weibull distribution," *Int. J. Eng. Adv. Technol.*, vol. 8, no. 6, pp. 1871–1874, 2019.
[13] J. Waewsak, C. Chancham, M. Landry, and Y. Gagnon, "An Analysis of Wind Speed Distribution at Thasala, Nakhon Si Thammarat, Thailand," *J. Sustain. Energy Environ.*, vol. 2, pp. 51–55, 2011.
[14] M. H. Al Buhairi, "A statistical analysis of wind speed data and an assessment of wind energy potential in Taiz-Yemen," *Assiut Univ. Bull. Environ. Res.*, vol. 9, no. 2, pp. 21–33, 2006.
[15] N. Sanusi, A. Zaharim, S. Mat, and K. Sopian, "A Weibull and finite mixture of the von Mises

distribution for wind analysis in Mersing, Malaysia," *Int. J. Green Energy*, vol. 14, no. 12, pp. 1057–1062, 2017.

[16] P. Wais, "A review of Weibull functions in wind sector," *Renew. Sustain. Energy Rev.*, vol. 70, no. 1, pp. 1099–1107, 2017.

[17] D. Indhumathy, C. V Seshaiah, and K. Sukkiramathi, "Estimation of Weibull Parameters for Wind speed calculation at Kanyakumari in India," *Int. J. Innov. Res. Sci. Eng. Technol.*, vol. 3, no. 1, pp. 8340–8345, 2014.

[18] N. Aries, S. M. Boudia, and H. Ounis, "Deep assessment of wind speed distribution models: A case study of four sites in Algeria," *Energy Convers. Manag.*, vol. 155, no. August 2017, pp. 78–90, 2018.

[19] F. G. Akgül, B. Şenoılu, and T. Arslan, "An alternative distribution to Weibull for modelling the wind speed data: Inverse Weibull distribution," *Energy Convers. Manag.*, vol. 114, no. 7, pp. 234–240, 2016.

[20] A. Zaharim, S. K. Najid, A. M. Razali, and K. Sopian, "Analyzing Malaysian wind speed data using statistical distribution," *4th IASME / WSEAS Int. Conf. ENERGY Environ. Anal.*, no. December, pp. 363–370, 2009.

[21] N. Masseran, A. M. Razali, K. Ibrahim, W. Z. W. Zin, and A. Zaharim, "On spatial analysis of wind energy potential in Malaysia," *WSEAS Trans. Math*, vol. 11, no. 6, pp. 467–477, 2012.

[22] N. Masseran, A. M. Razali, K. Ibrahim, A. Zaharim, and K. Sopian, "The Probability Distribution Model of Wind Speed over East Malaysia," *Res. J. Appl. Sci. Eng. Technol.*, vol. 6, no. 10, pp. 1774–1779, 2013.

[23] N. Sanusi, A. Zaharim, and S. Mat, "Wind energy potential: A case study of Mersing, Malaysia," *ARPN J. Eng. Appl. Sci.*, vol. 11, no. 12, pp. 7712–7716, 2016.

[24] T. B. M. J. Ouarda *et al.*, "Probability distributions of wind speed in the UAE," *Energy Convers. Manag.*, vol. 93, pp. 414–434, 2015.

[25] D. Derome, H. Razali, A. Fazlizan, A. Jedi, and K. Purvis-roberts, "Determination of Optimal Time-Average Wind Speed Data in the Southern Part of Malaysia," *Baghdad Science Journal*, vol. 19, no. 5, pp. 1111–1122, 2022.

[26] E. Dokur and M. Kurban, "Wind Speed Potential Analysis Based on Weibull Distribution," *Balk. J. Elec. Comp. Eng.*, vol. 3, no. 4, pp. 231–235, 2015.

[27] C. Jung, D. Schindler, J. Laible, and A. Buchholz, "Introducing a system of wind speed distributions for modeling properties of wind speed regimes around the world," *Energy Convers. Manag.*, vol. 144, pp. 181–192, 2017.

[28] M. Z. Ibrahim, K. H. Yong, M. Ismail, A. Albani, and A. M. Muzathik, "Wind characteristics and GIS-based spatial wind mapping study in Malaysia," *J. Sust. Sci. Manag.*, vol. 9, no. 2, pp. 1–20, 2014.

[29] R. I. Harris and N. J. Cook, "The parent wind speed distribution: Why Weibull?," *J. Wind Eng. Ind. Aerodyn.*, vol. 131, pp. 72–87, 2014.

[30] G. Erdemir, A. T. Zengin, and T. C. Akinci, "Short-term wind speed forecasting system using deep learning for wind turbine applications," *Int. J. Electr. Comput. Eng.*, vol. 10, no. 6, pp. 5779–5784, 2020.

[31] A. Belhamadia, M. Mansor, and M. A. Younis, "A study on wind and solar energy potentials in Malaysia," *Int. J. Renew. Energy Res.*, vol. 4, no. 4, pp. 1042–1048, 2014.

[32] N. Sanusi and A. Zaharim, "An Initial Study on Wind Speed in Mersing, Malaysia," in *Computer Applications in Environmental Sciences and Renewable Energy*, 2015, vol. 37, no. 1, pp. 110–113.

[33] A. K. Azad, M. G. Rasul, and T. Yusaf, "Statistical diagnosis of the best Weibull methods for wind power assessment for agricultural applications," *Energies*, vol. 7, no. 5, pp. 3056–3085, 2014.

[34] S. Glen, "Anderson-Darling Test & Statistic: Definition, Examples - Statistics How To." [Online]. Available: https://www.statisticshowto.datasciencecentral.com/anderson-darling-test/.

[Accessed: 17-Dec-2019].

[35] D. K. Kidmo, R. Danwe, S. Y. Doka, and N. Djongyang, "Statistical analysis of wind speed distribution based on six Weibull Methods for wind power evaluation in Garoua, Cameroon," *J. Renew. Energies*, vol. 18, pp. 105–125, 2015.

[36] D. Derome, H. Razali, A. Fazlizan, and A. Jedi, "Application of a New Method 'AGM' to Estimate Weibull Parameters for Low Wind Speed," *Int. J. Renew. Energy Res.*, vol. 12, no. 3, pp. 1362–1371, 2022.

[37] M. H. Zhang, "Wind Resource Assessment and Micro-Siting: Science and Engineering," in *Wind Resource Assessment and Micro-Siting: Science and Engineering*, First Edit., John Wiley & Sons Singapore Pte Ltd, 2015, p. 269.

The effect of trailing-edge wedge-tails on the aerodynamic characteristics of wind turbine airfoil

Asmail A M Abdalkarem[1], Ahmad Fazlizan Abdullah[1]*, Kamaruzzaman Sopian[1], Lim Chin Haw[1], Wan Khairul Muzammil[2] and Kok Hoe Wong[3]

[1]Solar Energy Research Institute (SERI), University Kebangsaan Malaysia, 43600 Bangi, Selangor, Malaysia.
[2]Energy Research Unit, Faculty of Engineering, University Malaysia Sabah, Jln. UMS, 88400 Kota Kinabalu, Malaysia.
[3]Department of Mechanical Engineering, Faculty of Engineering and Physical Sciences, University of Southampton Malaysia, 79200 Iskandar Puteri, Malaysia.

*Corresponding Author: a.fazlizan@ukm.edu.my

Abstract. Passive flow control devices have been proposed by many researchers for suppressing dynamic stall and enhancing the aerodynamic performance of turbomachinery including wind turbines. These technologies are usually cheap, easier to implement, and no additional energy needed. In the present study, the effects of adding trailing edge wedge tail (WT)on the aerodynamic characteristics of a static airfoil have been evaluated. The aerodynamic behavior of 2D-NACA0021 airfoil equipped with WT have been studied numerically using the commercial code ANSYS FLUENT to solve the Reynolds-Averaged Navier–Stokes equation (RANS). The effect of varying heights and lengths to height ratios (L/H) of WT have been studied. The results showed that the aerodynamic efficiency of NACA0021 airfoil influenced significantly by height of WTs comparing to slightly influence by length at L/H<1, and the maximum improvements of the lift and lift-to-drag ratio of the airfoil up to more than +38% and 28.8% respectively at optimum height 1.75%c and length 1%c compared to baseline airfoil case and suggested that to add trailing WTs as a retrofit to existing wind turbines.

1. Introduction

In recent decades horizontal axis wind turbines (HAWTs) received main attention from scientists and industry worldwide and its efficiency is better improved due to the benefits from the enormous investment to tackle many deficiencies [1]. Compared with HAWTs, vertical axis wind turbines (VAWTs) have been receiving renewed attention for several advantages such as: [1] [2] Ability to withstand high turbulence and better performance in complex urban areas and operate at low wind speeds, Omni-directional capability, low-cost of (manufacturing, installation and maintenance). Even though VAWT have many advantages, it is also confronting some challenges such as low power generating due to dynamic stall, blade wake interactions and difficults in self-starting. Therefore, many flow control techniques have been proposed for suppressing dynamic stall and enhancing the power coefficient of turbomachinery, including the VAWT.

Trailing edge (TE) flaps including Gurney flaps have been studied by many researchers as passive solutions for power augmentation of VAWTs. First TE modification known as Gurney flap (GF) attributed to the American inventor Dan Gurney. Gurney flap is flat plate added to the TE of the airfoil perpendicular to the chord line on its lower side. Trailing edge modification first proven by Liebeck [3]. Liebeck observed that changing the Kutta condition location producing additional circulation thus, enhancing lift and drag coefficients generated with the downward turning of the flow near the TE. Liebeck observed that GF having a height 1.25% of the chord length (c) increases the lift with only a small penalty of drag coefficient and the penalty of drag rises rapidly for flap heights larger than 2%c which is supported later by several researchers [4] [5] [6].

Lift enhancement modification produces high lift however, the drag is also increased due to the flow separation downstream of the TE. Different drag reduction modifications have been proposed by many authors such as [7] Neuhart developed a serrated GF added to NACA0012 airfoil to eliminate the drag penalty. [8] Bechert studied different GF modifications including GF with slite and GF with holes and later by [9] and [10]. The results show that, small size of holes and slites limited the effectiveness of GF and with higher porosity the mechanical stiffness and stability of the GF decreased. First attempt to investigate and utilize the potential of wedge flap (WF) was by Boyd [11]. He conclude that the preferred height is 0.5%c a placed at 0-1%c from TE. [12] Bloy tested and compared C_L/C_D for five flap shapes with the same length attached to NACA 5414 airfoil named: 90° WF, 90° GF, 45° WF, 45° GF, and square section. The results show that WFs produced the highest C_L/C_D ratio throughout all studied flaps. Timmer [13] investigated the effects of adding GFs and WFs to DUT 93-W-210 airfoil on the aerodynamic performance. The results revealed that WF and GF show similar TE devices characteristics at 1%c. Mohammadi [14] studied the aerodynamic performance of DU 91-W2-250 airfoil equipped with TE curved wedge flap, TE wedge, and GF. They concluded that both types of WFs have C_L/C_D exceeded the GF with the same height of 1%c. In both WFs drag coefficient decreased compared to GF which is attributed to disappears of separation bubbles at the upstream of the TE.

Previous literatures show different attempts by many researchers to reduce the drag coefficient, in order to improve lift to drag ratio for this purpose different devices have been equipped on only pressure side of the airfoil. However, dynamic stall occurs periodically on both outer and inner sides of the symmetrical blades of straight bladed-VAWTs [15]. This study aimed to preliminary assessment of 2D NACA 0021 airfoil equipped with two sides wedge flap named wedge tail and analysed its ability to improve the aerodynamic performance numerically using ANSYS FLUENT software. Proper size of WT dimensions (height and length) at maximum aerodynamic performance have been studied thus, VAWT can be modified based on optimum wedge tail sized.

2. Computational methodology

In current study, steady CFD simulation using ANSYS solver based on the finite volume method employed to predict the fluid flow pattern by solving numerically set of governing mathematical equations.

2.1. Airfoil geometry

The well-known symmetrical NACA0021 airfoil commonly used for VAWT applications. It has relatively large thickness and improves wind turbine performance. In present study 4-digit NACA 0021 has been selected with chord length (c)140mm, which has been studied numerically and experimentally by some researchers in different works [16][17] [18]. The effect of different WT dimensions (height and length) was simulated and compared to the baseline airfoil, as following:

- *Effects of WT height*

The geometry of the NACA0021 airfoil equipped with WT is shown in Figure 1, where H refers to the height of the half of the WT perpendicular to the chord line and L is the length of WT extended with an airfoil surface, both halves of WT are symmetrical. In this section the dimensions of WT selected to be at the same dimension L=H. Different WT heights were studied reaching to the optimum height at maximum (C_L/C_D). Both the length (L) and height (H) of the flap was defined relatively to the chord length.

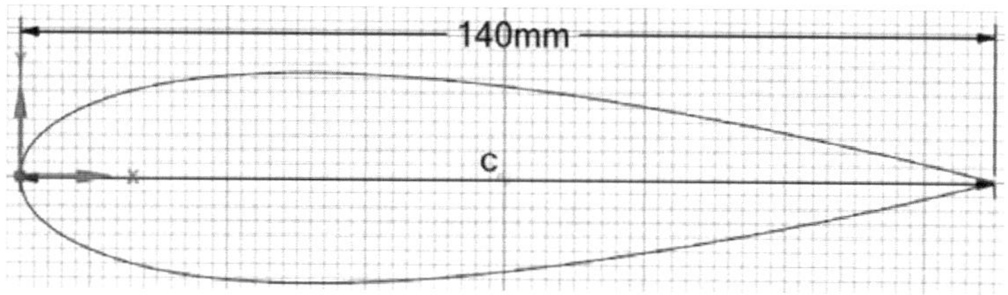

(a)

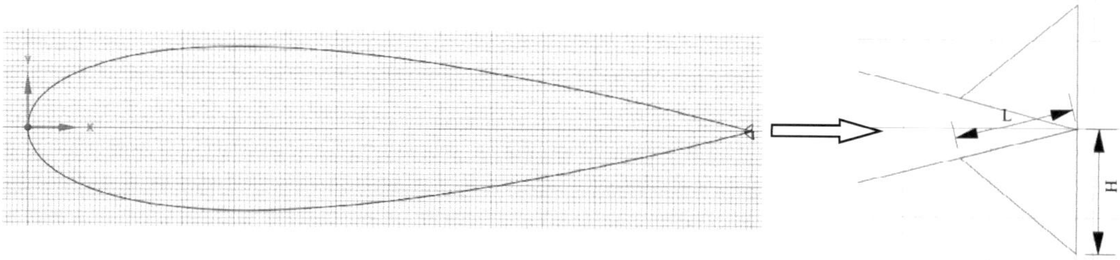

(b)

Figure 1. Airfoil geometry of NACA 0021 (a) baseline airfoil (b) Airfoil with WT.

- *Effects of WT length*

In current study the influence of various L/H ratios have been studied L/H <1 and L/H >1 at constant optimum height from previous section on the performance of static NACA 0021.

2.2. Computational equations and numerical solutions

The governing equations of CFD are the Reynolds-averaged Navier–Stoke (RANS) equations. These equations were derived from the conversation law of the physical properties of working fluid. The shear stress transport (SST) k-ω turbulent model has two equations were developed by Menter [19]. The proposed SST k-ω turbulence model considers the influence of turbulent shear stress transportation, also has a better computational accuracy to capture the adverse pressure gradient flows. In this study numerical simulations were performed using ANSYS FLUENT commercial CFD software, to solve the RANS equations, the solution methods were selected as shown in table 1.

Table 1. Solution method.

Pressure-velocity coupling	Coupled
Gradient	Least squares cell based
Pressure	Second order upwind
Momentum	Second order upwind
Turbulent kinetic energy (k)	Second order upwind
Specific dissipation rate (ω)	Second order upwind

Different studies have been done by many researchers to investigate the ability of SST k-ω to capture the characteristics of dynamic stall. For example Rezaeiha [20] performed a comparing between seven turbulence models commonly used to optimize the highest accurate model for CFD simulations of VAWT. The extensive analysis shows that only the SST models including SST k-ω, has the capability of exhibiting reasonable agreement with all the experimental cases tested and can predict the power coefficient more accurately at wide range of tip speed ratio (TSR). Thus, the turbulence model employed in this study is the shear stress transport (SST) k-ω.

2.3. Domain meshing and boundary conditions

In the CFD simulation of a static airfoil, The Reynolds number is Re =180,000 based on the airfoil chord length is 0.14 m and the Mach number is 0.06, i.e. the flow is incompressible. A C-type with size of computational domain in the range of 10 to 20 length /chord have been adopted by many researchers such as [21] [22]The distance of the outlet and inlet boundaries away from the 1/4 of the airfoil chord length is 22.5c and 12.5c, respectively and the bottom and top boundaries are 21.75c away from the TE. Table 2 shows boundary conditions for CFD calculation.

Table 2. Boundary conditions for CFD simulations.

Boundary	**Type**
Inlet (Top, Left, Bottom)	Velocity inlet, 20 m/s
Right	Pressure outlet, atmospheric pressure value
Airfoil and the WT surfaces	No-slip wall

2.4. Grid generation and sensitivity

Different mesh properties contribute together to improve the quality of the mesh which is leading to the increment in the accuracy of calculation results avoids the waste of computational time. In this study A C-type structured grid was adopted, and the grid distribution near the wall where the boundary layer located has been refined to capture accurately the physics of the flow around the airfoil. The structured mesh in the far field (the wake region) also generated. **Error! Reference source not found.** presents the airfoil section, whole of computational domain and a zoomed view of the leading and TE-WT including different ratios (L/H=1 case, L/H<1 and L/H>1).

Different types of grids were used in present simulation to check the sensitivity of the results on the grids. The details of the grids studied at Re=180,000. **Error! Reference source not found.** shows the variations of C_L with range of AOA from 0° to 20° for different grid distributions G1, G2, G3 and G4 with number of cells 27,300, 64,000, 108,000 and 130,900 respectively.

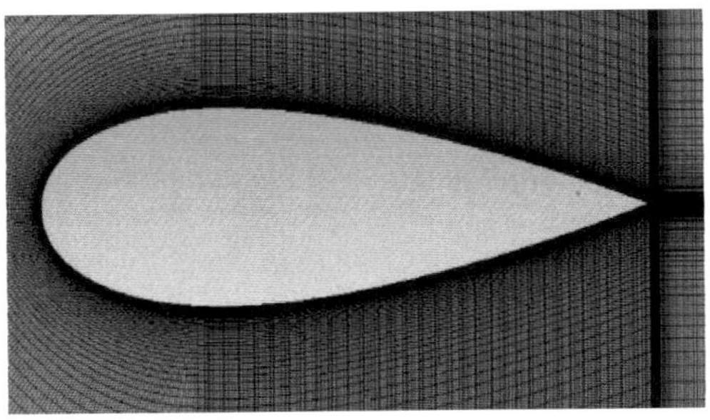

(a)

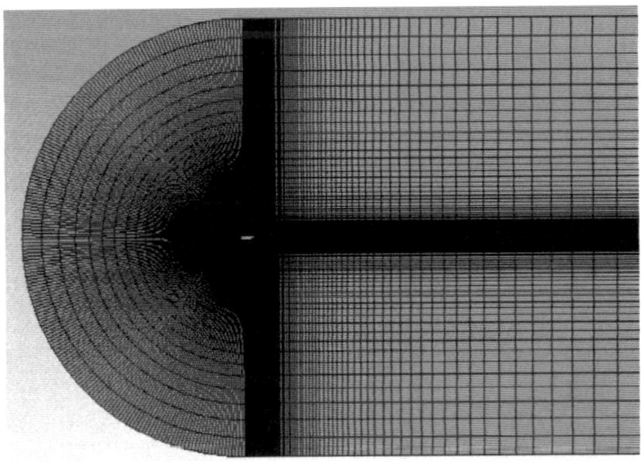

(b)

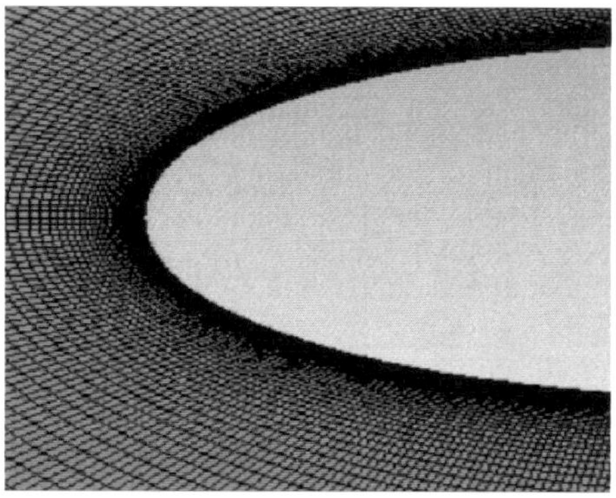

(c)

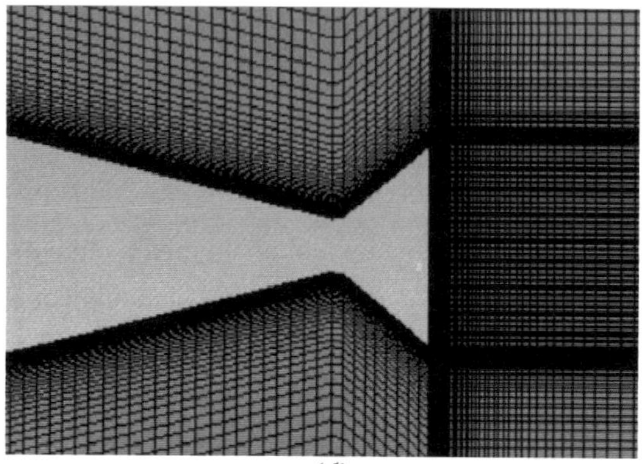

(d)

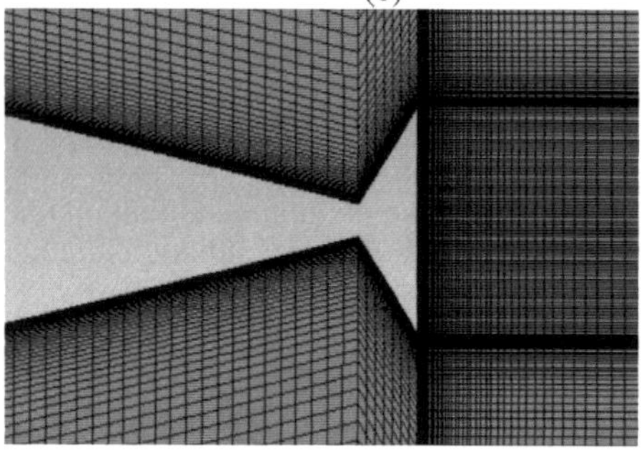

(e)

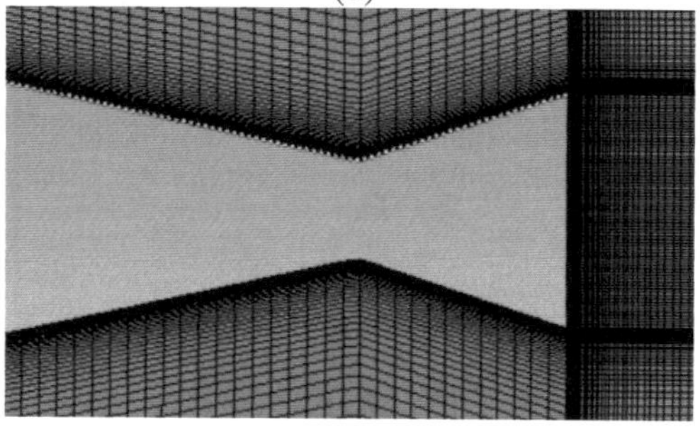

(f)

Figure 2. Details of computational grid (a) Airfoil (b) C-mesh computational domain (c) Near the leading-edge (d) Near TE around WT at L/H=1 (e) Near TE around WT at L/H<1 (f) Near the TE around WT at L/H>1.

The coefficient of lift increases with grid refinement from G1 to G3 and G4 shows no more increases in lift coefficient with higher mesh density, which is indicates that the CFD simulation results are independent on the cell numbers for grid selected. Therefore, the G3 with number of cells 108,000 for the case of baseline airfoil was adopted in this study. While the grid cells for the airfoil equipped with WT is about 117,800. As shown in **Error! Reference source not found.**, the 800 nodes along the chord line on the lower and upper airfoil surfaces are meshed with higher density near the trailing and leading-edge regions. The non-dimensional distance y^+ values of the first grid points kept ≈ 1.

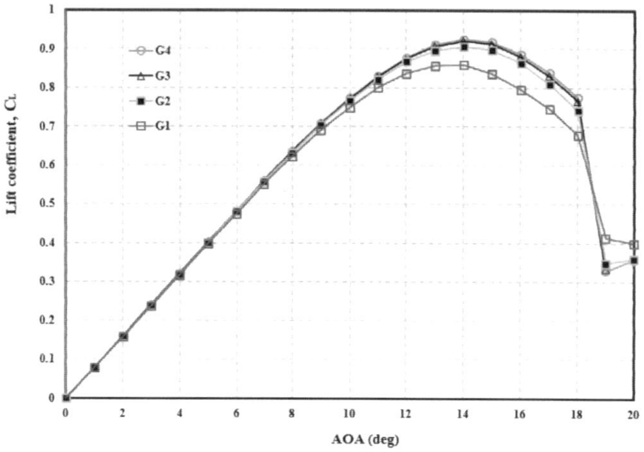

Figure 3. Verification of CFD simulation for different grid distributions for lift effects in the case of baseline static airfoil.

IOP Conf. Series: Materials Science and Engineering 1278 (2023) 012011 doi:10.1088/1757-899X/1278/1/012011

3. Results and discussion

3.1. Validation study

The validation of the present CFD simulation was performed by comparing lift and drag coefficients for the static airfoil with measurements from previous wind tunnel experiments and CFD simulations for the same airfoil details (chord length, Reynolds number, NACAxx series) with current study. The graph in Figure 4 presents computational and experimental data for the NACA 0021 at ranges of AOAs from 0° to 20°. The literature data was taken out of some previous works, the wind tunnel experiment by Holst [17] and CFD simulation work recently done by Balduzzi [23]. Despite the varying of different wind-tunnel configurations and experimental conditions, most of the literatures show a suitable matching to the present study in terms of the slope. The drag values of current CFD study show very good agreement with the numerical literature data before static stall. The general behavior of lift in the present study shows good agreement with work of Balduzzi despite of slightly underestimated of maximum C_L by CFD. This might be attributed to the varying of domain topology and the turbulent model was not fully turbulent, and the present simulations have assumed fully turbulent flow.

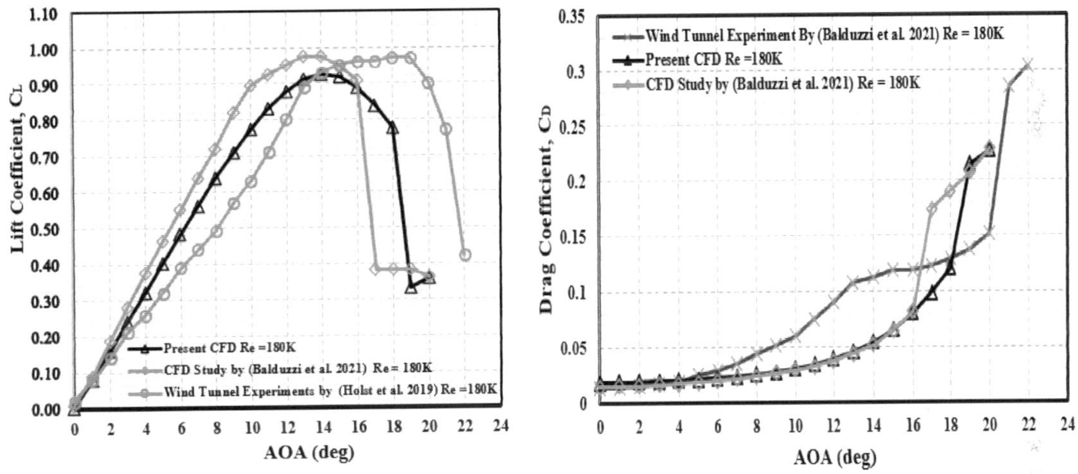

(a) (b)

Figure 4. Validation of static polars (NACA 0021 profile) against literature data at Re~180K **(a)** Lift **(b)** Drag.

3.2. Trailing Edge Structure

The impacts of WT on the TE structure are shown in Figure 5. Figure 5 shows Streamlines, velocity and pressure distributions of NACA 0021 airfoil employed with different WT dimensions at optimum height (H=1.75%c) and various lengths L/H<1(L=1%c), L/H=1 and L/H>1 (L= 3%c) at AOAs 0°, 9° and 13° in comparison with flow visualization of baseline airfoil. Adding WTs to baseline airfoil change the Kutta condition as shown in Figure 5 at AOA=0 the streamline structure shows two symmetrical counter rotating vortices in the downstream zone of the WT in additional to presence of two vortices in the upstream zones of WTs. At AOA>0 and with presence of the WT, the pressure on the pressure surface increases as shown in pressure contours at AOA=9 in comparison with baseline airfoil case, in addition, existence of the vortices which is suppress the separation point near the TE. thus, the lift increases. As the length of the two-sides of WT increases upstream at L/H>1 the lift and drag decreased, this might be due to the decreases in the size and shifting in the location of the upstream vortices.

IOP Conf. Series: Materials Science and Engineering 1278 (2023) 012011 doi:10.1088/1757-899X/1278/1/012011

The WT, L/H<1 shows higher circulation in the upstream zone comparing to WT, L/H>1 this might be due to the higher lift and drag produced by WT, L/H<1 while WT, L/H=1 shows smoothly flow deflection which is producing less lift but considerably less drag with enhancing lift to drag ratio. With more increasing of AOAs, the airfoil experiences complete flow separation and there is a sudden decrease in the lift referring the occurrence of dynamic stall with large turbulent wake behind the TE.

Flow Visualization

Streamlines	Velocity contours	Pressure contours

AOA=0

Baseline airfoil / WT, L/H<1 / WT, L/H=1 / WT, L/H>1

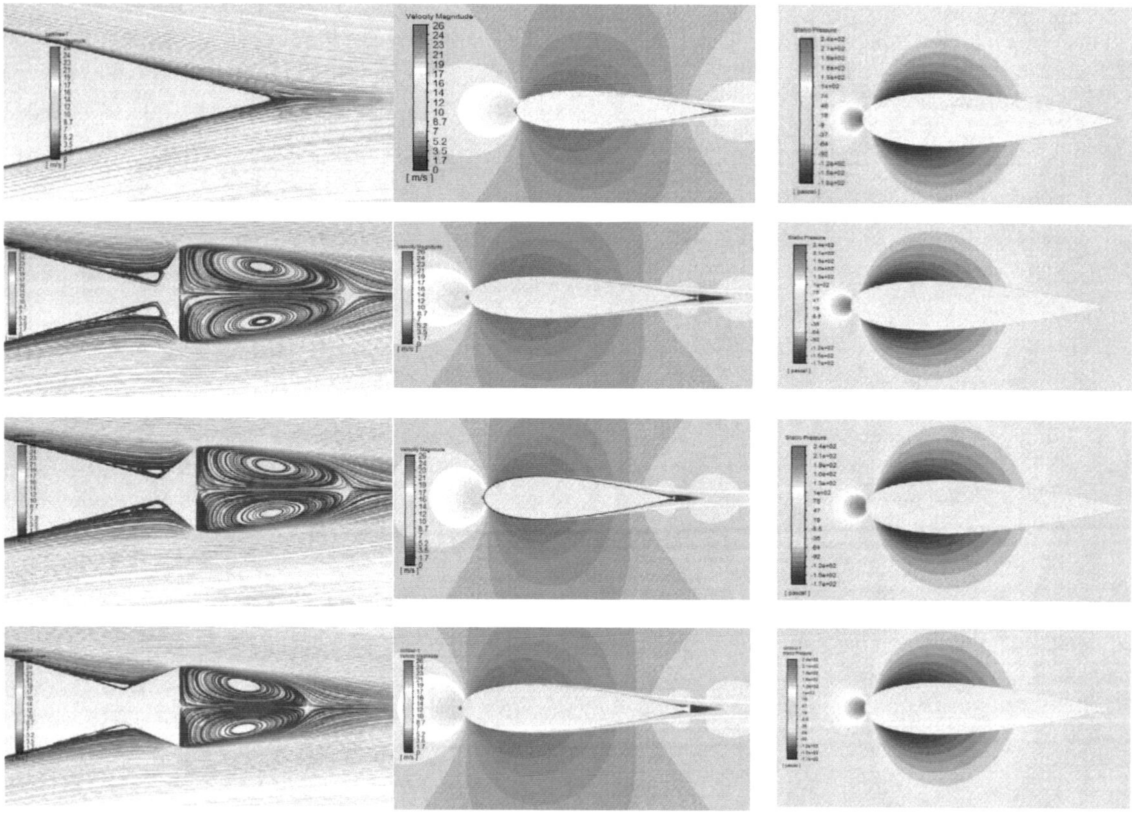

Flow Visualization

Streamlines	Velocity contours	Pressure contours

AOA=9

Baseline

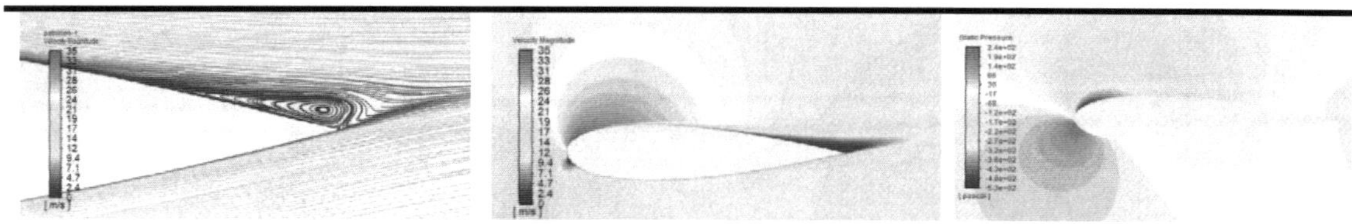

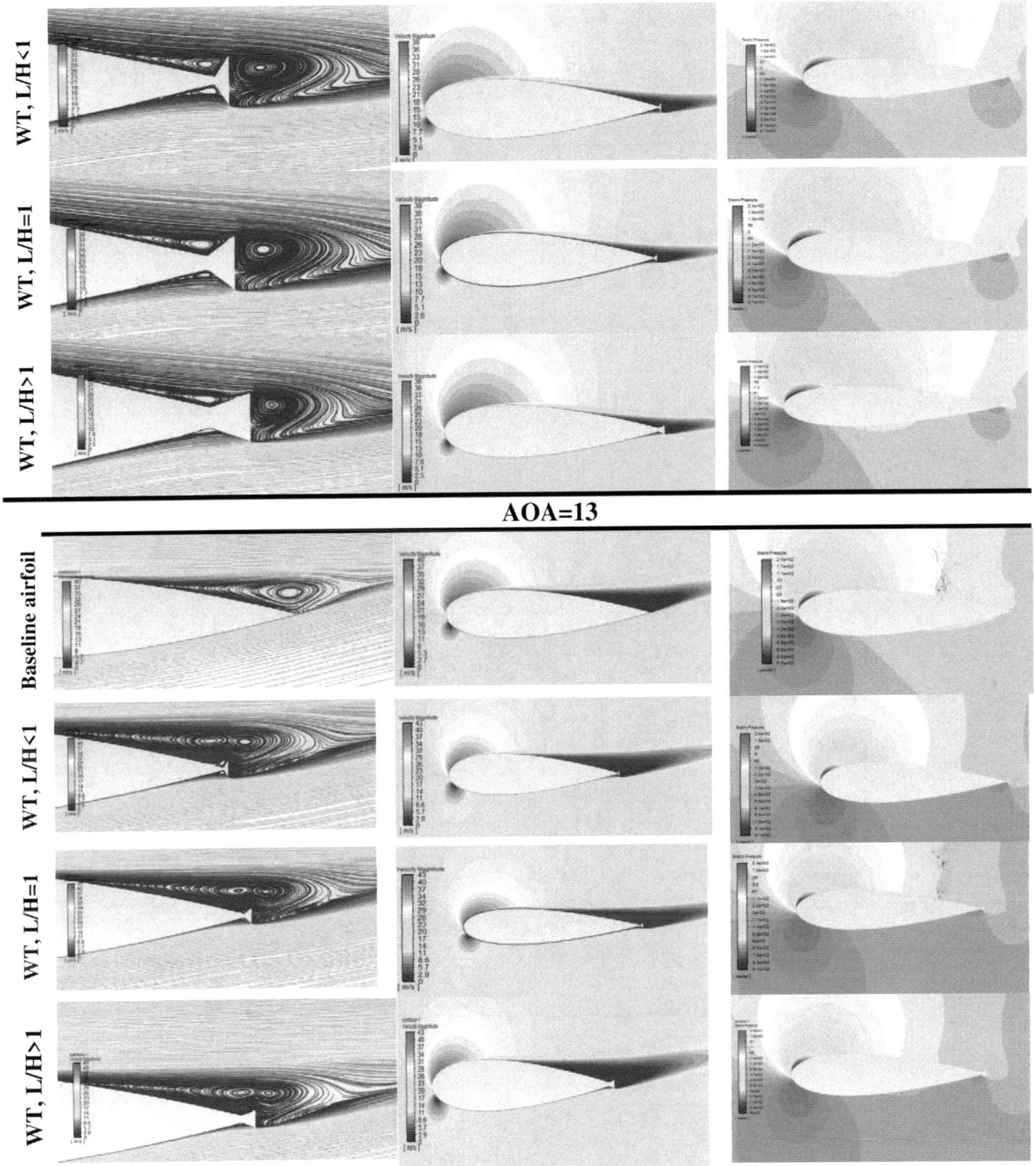

Figure 5. Streamlines, pressure, and velocity distributions for various WT cases at different AOAs in comparison with baseline NACA 0021 airfoil.

3.3. Effect of WT height

The influence of the wedge height added to the NACA 0021 airfoil was studied. In this section both wedge dimensions L and H have the same length. **Error! Reference source not found.**a demonstrates the increments in the lift coefficient with increasing of WT height. Wedge tail shifts the C_L– AOA curve to lift by 1° in comparison with the baseline airfoil case. The values of the maximum C_L

IOP Conf. Series: Materials Science and Engineering 1278 (2023) 012011 doi:10.1088/1757-899X/1278/1/012011

increased by 12.93%,23.98%, 36.92%, 40.37% and 53.05% for the wedge heights 0.5%c, 1%c, 1.75%c, 2%c and 3%c, respectively. Lift coefficient curve versus AOA also shows that the stall angle decreases from 18° of the baseline airfoil case to 16° of the case with a 3%c of the wedge height.

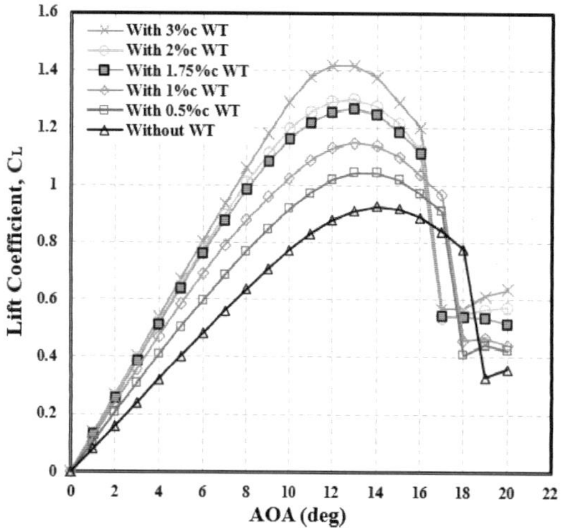

(a)

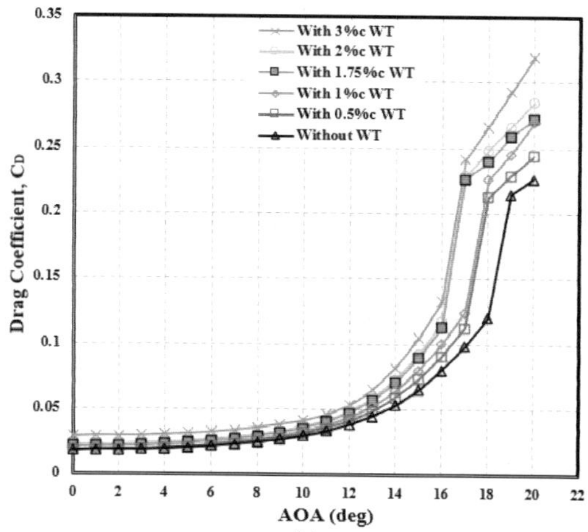

(b)

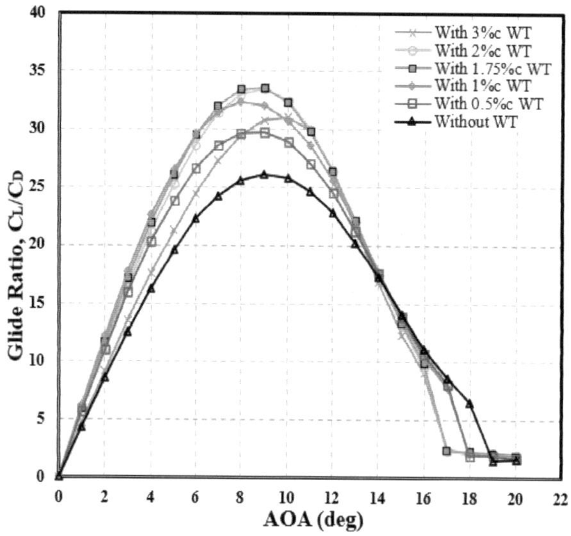

(c)

Figure 6. Effect of WT height on the aerodynamic characteristics of the NACA 0021 airfoil. **(a)** Lift. **(b)** Drag. **(c)** Glide ratio.

The effects on drag coefficients using various wedge tail heights illustrated by Figure 6b. Obviously, as the AOA increases the C_D increases as well. At low AOA the drag slightly increased while, at moderate to high of AOAs, there is a large drag penalty associated with the WTs. It is also clear that as the wedge height increases the drag significantly increase. Figure 6c illustrates the increases in glide ratio with AOA at different WT heights in comparison with baseline airfoil. At low to moderate AOAs the lift coefficient increases with small of drag penalty resulting to rapid increase in glide ratio, while at moderate-to-high AOAs there is large drag penalty leading to decreases in glide ratio.

Increments in the maximum values of glide ratio at different wedge heights presented in Figure 7. Glide ratio increases with wedge heights of 0.5%c, 1%c, 1.75%c, while wedge with 2%c and 3%c heights show reduction in glide ratio due to large penalty in drag this might be attributed to the higher increase of WT height which is exceeded the height of boundary layer thickness. Therefore, the optimum Wedge height is H=1.75%c (which is equivalent to 3.5%c for whole WT height) with maximum increment in the values of glide ratio is 28.60 % compared to the baseline airfoil case.

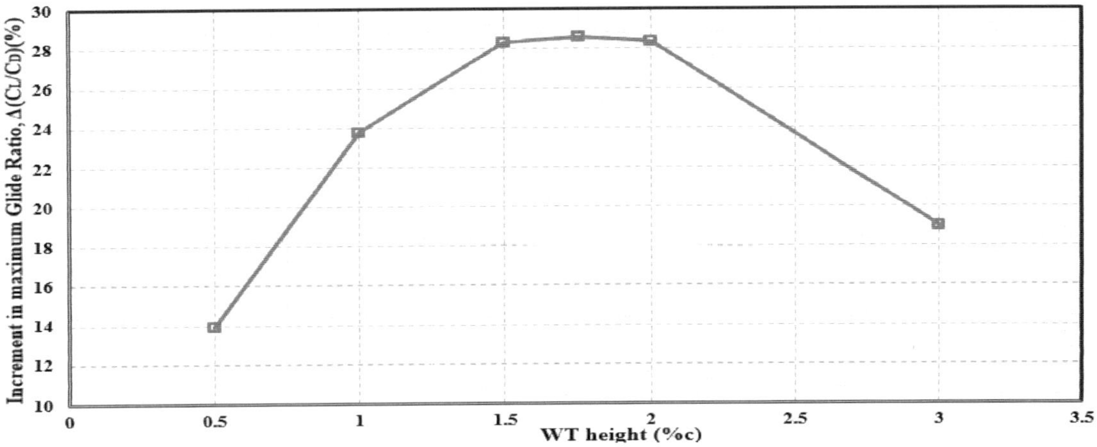

Figure 7. Increments in the maximum values of glide ratios at different WT heights in comparison with the baseline airfoil.

3.4. *Effect of WT length*

In the present section the effect of different WT lengths at optimum constant height (resulted from previous section) on the performance of static NACA 0021 will performed. Various L/H ratios have been studied at L/H<1 and L/H>1 as shown in Figure 8 in comparison with optimum characteristics of WT at heights of 1.75%c (L/H=1) and the results presented in Figure 8, Figure 9 and Table 3. The C_L-AOA curve shows positive increments of maximum lift at L/H<1 while at L/H>1 the increment of maximum C_L values became negative, this might be attributed to the increases in penalty of drag due to extension in the length at upstream.

The lift coefficient curve shows highest lift value at L/H =0.285 (L=0.5, H=1.75%c) accompanied with highest penalty of drag resulting to the low values of glide ratio as shown in Figure 8c , this might be happened due to the vortex tip losses occurs at TE tips. As the L/H ratio increases at pressure side of the airfoil the lift decreases and the drag also decrease but the glide ratio increased reaching to its maximum increment value of glide ratio with 0.16%c in comparison to the L/H=1 case at optimum height and length L/H=0.571(L=1%c, H=1.75%c) before its significantly decrease at L/H>1.

Table 3. Increments in the maximum lift coefficient and glide ratio values at optimum constant WT height H= 1.75%c (2.45 mm) and different WT lengths in comparison with the L/H=1 case.

TE-WT at H=1.75%c, c=140mm				
WT length (L) based on chord length (%)	WT length L(mm)	L/H ratio	Increments in maximum C_L (%)	Increment in maximum Glide Ratio, C_L/C_D (%)
0.5%c	0.7	0.285	1.84	-0.044
1%c	1.4	0.571	0.85	0.16
1.75%c	2.45	1	0	0
3%c	4.2	1.714	-2.06	-0.42

The effect of wedge length at L/H<1 shows slightly improvements on the aerodynamic performance compared to L/H=1, while L/H>1shows higher decrement in glide ratio.

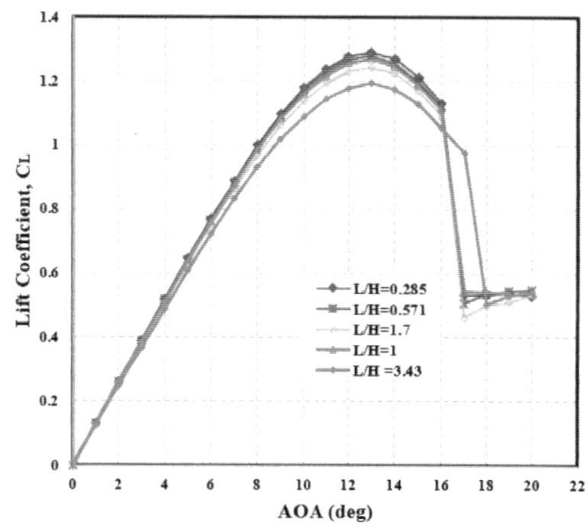

(a)

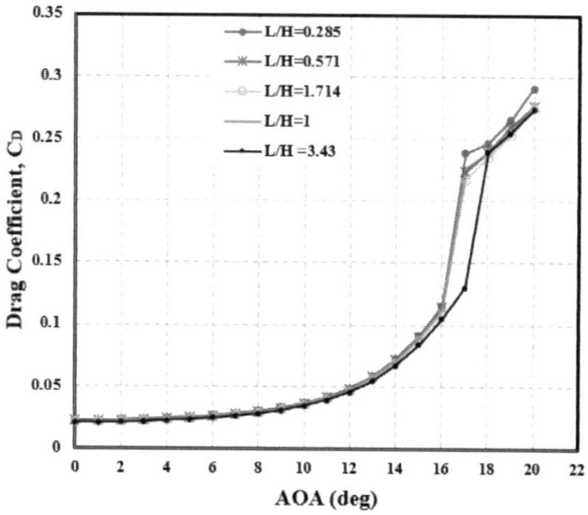

(b)

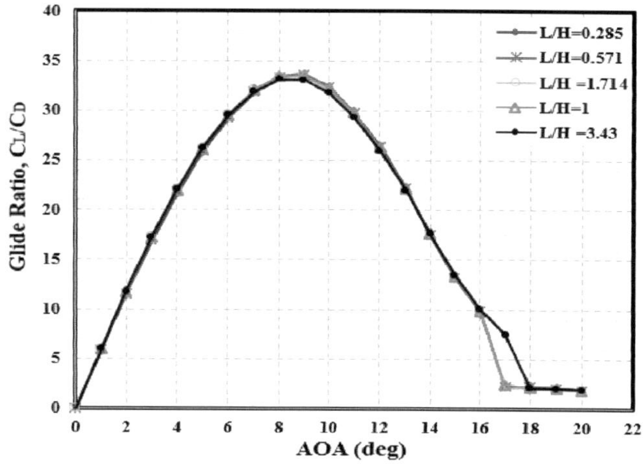

(c)

Figure 8. Effect of WT length on the aerodynamic characteristics of the NACA 0021 airfoil. (a) Lift (b) Drag (c) Glide ratio.

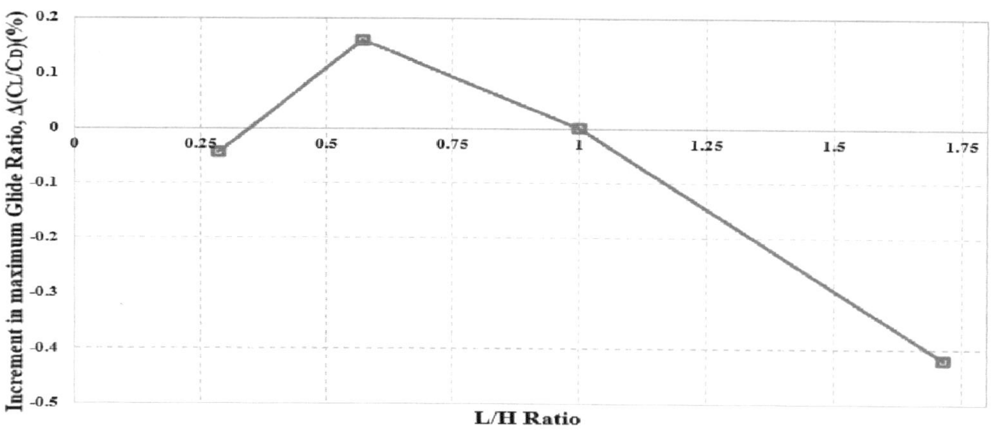

Figure 9. Increment in the values of maximum glide ratio at different WT lengths in comparison with the L/H=1 case.

3.5. Optimum aerodynamic characteristics of TE-WT

Maximum aerodynamic characteristic for different WT cases at L/H=1 and L/H<1 compared to baseline airfoil case are shown in Figure 10. In Figure 10*a* the WT at L/H<1 *shows maximum aerodynamic performance at optimum height and length* (1.75%c and 1%c) respectively with maximum lift coefficient value of 1.28 at AOA = 13 and total increment value about 38.08 % in comparison of baseline airfoil case. Figure 10b shows reduction in drag of L/H=1 case at AOA<10 before its increases at higher AOAs in comparison to L/H<1 case. The aerodynamic efficiency for WT with L/H=0.571 (L=1%c) in Figure 10*c* shows the highest maximum increment values of glide ratio with 28.80 % at AOA = 9 in comparison to the baseline airfoil case.

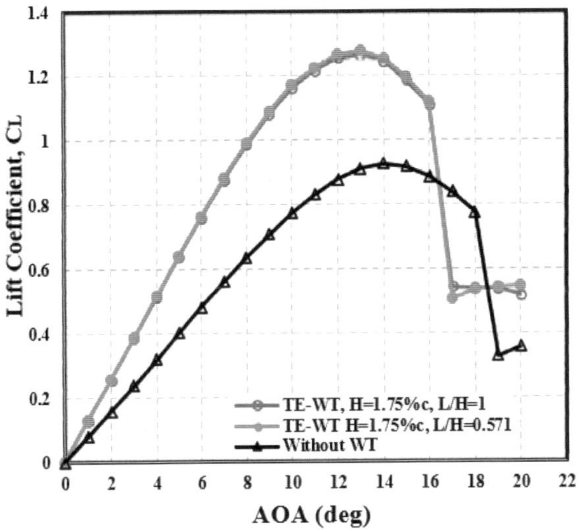

(a)

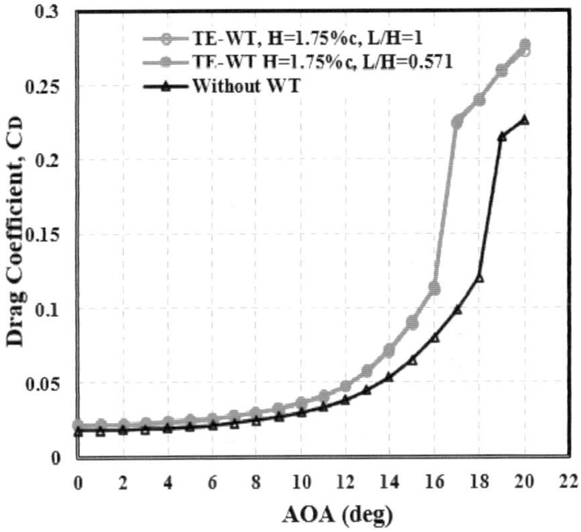

(b)

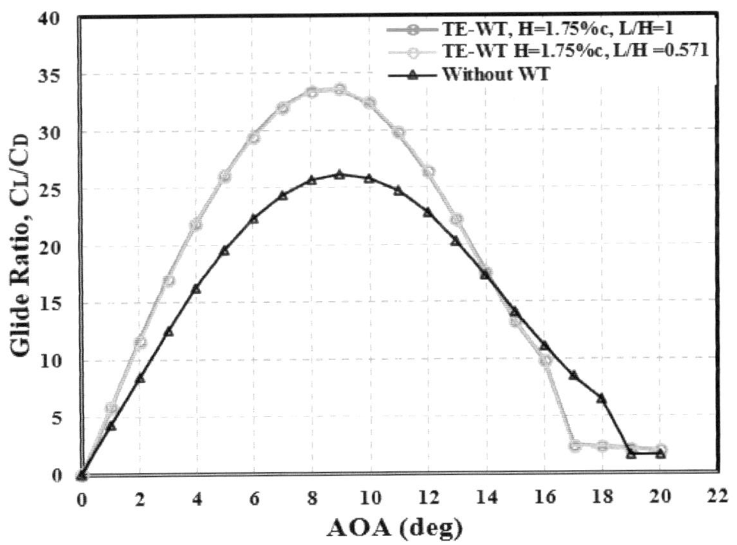

(c)

Figure 10. Optimum aerodynamic characteristics for TE-WT in comparison with the clean NACA 0021 airfoil (a) Lift (b) Drag (c) Aerodynamic efficiency.

4. Conclusions

In this study, CFD simulations was performed on two-dimensional NACA 0021 airfoil with TE-WT at Re=180,000 to solve the steady Reynolds Averaged Navier Stokes equation (RANS) at wide range of

AOAs. Various cases have been considered to investigate the impact of WT height and the effect of different (L/H) ratios on the aerodynamic characteristics and the flow over the airfoil. Results show that, adding WT at TE of NACA 0021airfoil could improves the aerodynamic characteristics in all studied cases compared to the baseline airfoil. The optimum WT height is 1.75%c at L/H=1 and the optimum WT length at 1.75%c height is 1%c achieving optimum dimensions at L/H<1 case with total maximum increment in glide ratio 28.80% at AOA=9° compared to the baseline airfoil. Wedge tails with L<1%c show noticeably decrease in maximum glide ratio this might be due to the vortices occurs at tip of blade or wing which is cause losses in overall performance. Wedge Tail height shows significantly effect on the aerodynamic performance while its length shows slightly effect.

5. References

[1] I. Paraschivoiu, "Wind Turbine Design: With Emphasis on Darrieus Concept," *Presses Internationales Polytechnique*. Canada, pp. 1–438, 2002.

[2] Q. Li, T. Maeda, Y. Kamada, Y. Hiromori, A. Nakai, and T. Kasuya, "Study on stall behavior of a straight-bladed vertical axis wind turbine with numerical and experimental investigations," *J. Wind Eng. Ind. Aerodyn.*, vol. **164**, no. April 2016, pp. 1–12, May 2017.

[3] R. H. Liebeck, "Design of Subsonic Airfoils for High Lift," *J. Aircr.*, vol. **15**, no. 9, pp. 547–561, Sep. 1978.

[4] Y. Li, J. Wang, and P. Zhang, "Effects of Gurney flaps on a NACA0012 airfoil," *Flow, Turbul. Combust.*, vol. **68,** no. 1, pp. 27–39, 2002.

[5] R. Myose, I. Heron, and M. Papadakis, "The Post-Stall Effect of Gurney Flaps on a NACA-0011 Airfoil," in *SAE Technical Papers*, 1996, no. January.

[6] D. Jeffrey, X. Zhang, and D. W. Hurst, "Some Aspects of the Aerodynamics of Gurney Flaps on a Double-Element Wing," *J. Fluids Eng.*, vol. 123, no. 1, pp. 99–104, Mar. 2001.

[7] D. H. Neuhart and I. C. Pendergraft, "Water tunnel study of Gurney Flaps," *NASA Tech. Memo.*, no. 4071, 1988.

[8] D. Bechert, R. Meyer, and W. Hage, "Drag reduction of airfoils with miniflaps - Can we learn from dragonflies?," in *Fluids 2000 Conference and Exhibit*, 2000.

[9] R. Meyer, W. Hage, D. W. Bechert, M. Schatz, and F. Thiele, "Drag Reduction on Gurney Flaps by Three-Dimensional Modifications," *J. Aircr.*, vol. 43, no. 1, pp. 132–140, Jan. 2006.

[10] T. Lee, "Aerodynamic characteristics of airfoil with perforated Gurney-type flaps," *J. Aircr.*, vol. 46, no. 2, pp. 542–548, 2009.

[11] J. A. Boyd, "Trailing edge device for an airfoil," *United States Pat.*, no. Patent Number: 4,542,868, pp. 1–6, 1985.

[12] A. W. Bloy, N. Tsioumanis, and N. T. Mellor, "Enhanced Aerofoil Performance Using Small Trailing-Edge Flaps," *J. Aircr.*, vol. 34, no. 4, pp. 569–571, Jul. 1997.

[13] W. A. Timmer and R. P. J. O. M. van Rooij, "Summary of the Delft University Wind Turbine Dedicated Airfoils," *J. Sol. Energy Eng.*, vol. 125, no. 4, pp. 488–496, Nov. 2003.

[14] M. Mohammadi, A. Doosttalab, and M. Doosttalab, "The effect of various gurney flap shapes

on the performance of wind turbine airfoils," *ASME Early Career Tech. Conf.*, vol. 11, no. November 2012, pp. 244–250, 2012.

[15] N. Fujisawa and S. Shibuya, "Observations of dynamic stall on Darrieus wind turbine blades," *J. Wind Eng. Ind. Aerodyn.*, vol. 89, no. 2, pp. 201–214, Feb. 2001.

[16] D. Holst *et al.*, "Static and Dynamic Analysis of a NACA 0021 Airfoil Section at Low Reynolds Numbers Based on Experiments and Computational Fluid Dynamics," *J. Eng. Gas Turbines Power*, vol. 141, no. 5, May 2019.

[17] D. Holst *et al.*, "Experimental Analysis of a NACA 0021 Airfoil Section Through 180-Deg Angle of Attack at Low Reynolds Numbers for Use in Wind Turbine Analysis," *J. Eng. Gas Turbines Power*, vol. 141, no. 4, Apr. 2019.

[18] F. Balduzzi *et al.*, "Combined Numerical and Experimental Study on the Use of Gurney Flaps for the Performance Enhancement of NACA0021 Airfoil in Static and Dynamic Conditions," *J. Eng. Gas Turbines Power*, vol. 143, no. 2, pp. 1–15, Feb. 2021.

[19] F. R. Menter, "Two-Equation Eddy-Viscosity Turbulence Models for Engineering Applications," *AIA A J.*, vol. 32, no. 8, 1994.

[20] A. Rezaeiha, H. Montazeri, and B. Blocken, "On the accuracy of turbulence models for CFD simulations of vertical axis wind turbines," *Energy*, vol. 180, pp. 838–857, 2019.

[21] F. Villalpando, M. Reggio, and A. Ilinca, "Prediction of ice accretion and anti-icing heating power on wind turbine blades using standard commercial software," *Energy*, vol. 114, pp. 1041–1052, Nov. 2016.

[22] J. Michna, K. Rogowski, G. Bangga, and M. O. L. Hansen, "Accuracy of the Gamma Re-Theta Transition Model for Simulating the DU-91-W2-250 Airfoil at High Reynolds Numbers," *Energies*, vol. 14, no. 24, p. 8224, Dec. 2021.

[23] F. Balduzzi *et al.*, "Combined Numerical and Experimental Study on the Use of Gurney Flaps for the Performance Enhancement of NACA0021 Airfoil in Static and Dynamic Conditions," *J. Eng. Gas Turbines Power*, vol. 143, no. 2, pp. 1–12, Feb. 2021.

NOMENCLATURE

Abbreviations

LE	Leading edge
TE	Trailing edge
GF	Gurney flap
WT	Wedge tail
WF	Wedge flap
SST	Shear stress transport
AOA	Angle of attack
G	Grid distribution
TSR	Tip speed ratio
HAWT	Horizontal axis wind turbine
VAWT	Vertical axis wind turbine
NACA	National Advisory Committee for Aeronautics
RANS	Reynolds averaged Navier-stokes

Symbols

c	Chord length
Re	Reynolds number
L	Length
H	Height
C_L	Lift coefficient
C_D	Drag coefficient
y+	Non-dimensional wall distance
U	Free stream velocity
C_L/C_D	Glide ratio

Decarbonization of tropical city using digital twin technology: case study of Bertam city

Nur Haziqah Mohamad Zaidi[1], Lim Chin Haw[1]*

[1]Solar Energy Research Institute, Universiti Kebangsaan Malaysia, 43600 Bangi, Selangor, Malaysia

*Corresponding author: chinhaw.lim@ukm.edu.my

Abstract. The urban population in Malaysia has increased to 77.7% in 2021, from 66% two decades ago, of which the current percentage shows Malaysia has more than 25 million city dwellers out of the 32.7 million total population. This statistic resulted in 80% of carbon emissions, roughly 200 million tCO^2e, being produced by the urban cities in Malaysia. To balance out the high number of carbon emissions from urban settings, our government has aimed to reduce 33% of carbon by 2030, aligning with the COP commitment by Malaysia in 2021, and later aiming for carbon neutrality by 2050. Carbon neutrality can be achieved by using digital twin technology, which can virtually represent any physical system, including a whole city to its real-world counterpart. Bertam, a city in Pulau Pinang is used as a case study in this study. A baseline simulation has been made of the digital twin city, resulting in 253 kWh/m^2 of total energy consumption in 2021, and 1.75 tonnes of carbon per capita. The recent applications of active and passive design strategies will be applied to the virtual city, to obtain its potential energy savings and carbon reduction data. At the end of the study, the digital twin technology will discover the city's potential energy savings and carbon reduction after the optimization.

1. Introduction

The impact of urbanization can be measured through the rising data of the total urban population in the world and nationwide. Recently, the UN stated more than 56% of the total world population lives in cities and the number will be expected to increase to 68% by 2050 [1]. In Malaysia, there is a rise of 11.7% of the urban population in the time span of two decades, from 66% in 2001 to 77.7% in 2021 [2]. Roughly, there are 25 million city dwellers, more than two-thirds inhabitants of Malaysia's total population which is 32.7 million. With the expansion of the urban population and areas, there will be an increase in activities in the urban areas, thus rising the city's carbon emissions and energy consumption. In 2020, Malaysia recorded an electrical energy consumption of 142.2 billion kWh, an increase of 7% in 5 years' time. From the number, 35% of the energy usage comes from the building sector [3]. A statistic in 2019 showed that 80% of carbon emissions in Malaysia came from urban settings, which accounted for 200 million tCO^2e [4]. Therefore, Malaysia has a record of 7.6 tonnes of carbon per capita in 2021. In a comparison of total carbon per capita, in tonnes, between other developing countries, in the same year, Malaysia is situated in the high carbon emission per capita ranking, slightly below China, 8.4[5]; and much higher than Thailand, 3.8 [6]; Indonesia, 2.2 [7]; and India, 1.9 [8].

Malaysia also relied on natural gas and coal as its main sources of power generation nationwide [9]. Electricity has become the main source of energy consumption in the country. In 2018, the electricity consumption of buildings sectors contributed to 41.5% of total energy consumption in the country [10].

IOP Conf. Series: Materials Science and Engineering 1278 (2023) 012012 doi:10.1088/1757-899X/1278/1/012012

This brings to the conclusion that urban settings and the building sector are the major contributors to the high carbon emissions and energy consumption rate. Thus, by tackling the city as the major contributor to the greenhouse gases issue, the consequences of the carbon emitted that leads to climate change can be slightly tempered in the future.

The Malaysian government has been committed to reducing the city's carbon emissions since the Conference of Parties (COP) 15 in Paris by publishing Low Carbon Cities Framework (LCCF) in 2011. The framework was implemented by the Malaysian Green Technology and Climate Change Centre (MGTC) to establish low-carbon zones in urban areas by 2030. It also consisted of an assessment system to perform a city's carbon reduction measures. In 2021, National Low Carbon Cities Masterplan (NLCCM) is developed by the Green Technology Application for the Development of Low Carbon Cities (GTALCC) to target a minimum of 33% carbon reduction in 33 major Malaysian cities by 2030 and thus, reaching carbon neutrality by 2050 and beyond [11]. Bertam city is under the Seberang Perai district, one of the targeted cities that is involved in this project. Hence, this study can be a benchmark for the local authorities on the potential carbon reduction and energy savings after the implementation of active and passive design strategies for the buildings in Bertam.

2. Methodology

2.1. Digital Twin Technology

The digital twin concept originated in 2002 at the University of Michigan's industry presentation by Michael Grieves. There are two systems, the physical system that exists and the virtual system containing all the data of the physical system. Therefore, a twinning or mirroring of the system remains in both real and virtual spaces [12]. Jones et. al in a study also stated that digital technology is a time saver, while still having high fidelity and accuracy of the data in its work of representing the real-time world virtually or vice versa [13]. slightly tempered in the future.

2.1.1 Intelligent Communities Lifecycle (ICL) for Bertam digital twin. Digital twin technology is beneficial to be implemented in the decarbonization of city planning, especially in estimating carbon emissions, calculating total energy consumption and decarbonizing a city. Intelligent Communities Lifecycle (ICL) is a digital twin technology created for decarbonizing any built environment at a scale of single buildings, campuses, communities, cities and a whole country. In order to achieve the goals of this study, a digital twin city that replicates the characteristics and behavior of Bertam city is created in the ICL software. A branch of the ICL, called Intelligent Community Design (iCD), is designed for creating sustainable master plan of a built environment. Therefore, it was chosen as a platform to create a digital twin for the city. Functioning as a plug-in in the SketchUp software, iCD created one central city model using OpenStreetMap, which has geodata underlying the digital twin models. Bertam is a city located at Peninsular Malaysia, in the state of Penang. It has a total gross floor area (GFA) of 3.66km^2 with an assumed population of 272 642 people. Figure 1 shows the proposed digital twin city of Bertam, created in the ICL software.

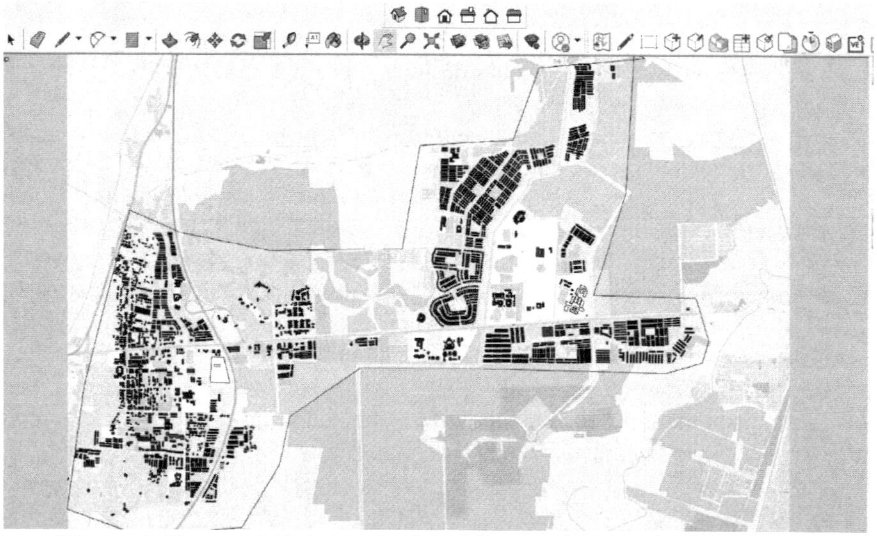

Figure 1. Proposed digital twin city of Bertam for sustainable master planning.

A simulation of energy performance if the digital twin city takes into account the location of the project and the data from the digital twin. The data of solar radiation, location, operating time of the building (24 hours, 12 hours, 9 hours), the glazing ratio (30%, 50%), construction materials and occupancy of the buildings were part of the data taken into consideration during the simulation. There are four planning scenarios of estimated energy consumption and carbon emissions for the digital twin city. Scenario 1 is the baseline masterplan where the virtual city model was assumed to represent the actual city. The parameters that significantly influence the energy behavior of the buildings were targeted such as the type of the building's envelopes and the HVAC system used. These parameters played important roles in determining the estimated energy demand of the buildings in the city. The master plan model was then used as a baseline to simulate the optimized parameters ranging from better thermal performance of the building envelopes to the HVAC system. Scenario 2 is the optimization of building envelopes where the heat coefficient transfer of building materials changed into lower amounts than the baseline master plan. Scenario 3 consists of the optimization of the HVAC system that used lower energy consumption than the previous system. Scenario 4 consisted of potential solar roof potential for the city of Bertam. A comparison between these four scenarios of buildings' energy consumption and carbon emissions was analyzed. Lastly, an analysis of the digital twin city's potential energy savings and carbon reduction was recorded.

3. Results & Discussions

3.1. Baseline data and post-optimization data of the Bertam digital twin city

The heat transfer coefficient (U-value in watt / m^2-Kelvin) of the roofs, ceilings, walls, windows, doors and floor materials majorly influenced the thermal condition of the spaces in the buildings [14]. For Scenario 1 - the baseline master plan, common construction types and building envelopes were directed to the models. Table 1 shows the comparison type and values of building envelopes and heat coefficient transfer for the city's baseline and optimized master plan in the year 2021.

Table 1. Heat coefficient transfer comparison for building envelopes in Scenario 1- baseline and Scenario 2 -lowering U-value master plan.

Building envelope	Type of building envelope	Heat transfer coefficient (U-value)	Type of building envelope	Heat transfer coefficient (U-value)
	Scenario 1		Scenario 2	
Roof	Sloping roof – domestic	3.38	19 mm. asphalt 13 mm. fibreboard 25 mm. air 75 mm. battens 10 mm. gypsum board	0.4
Ceiling	12mm plaster ceiling with TG boards on 400 joist center	1.69	8-in. lightweight concrete deck with false ceiling	0.81
Wall				
-Internal	115mm single-leaf brick (plastered on both sides)	1.97	13 mm. lightweight plaster 100mm lightweight concrete block 13 mm. lightweight plaster	1.05
-External	Brickwork single-leaf construction with dense plaster	2.18	2 in. insulation with 8 in. common brick	0.56
Windows				
-Internal	Single-glazed windows – domestic	3.3	Low-emission double-glazing - domestic	1.74
-External	Single-glazed windows – domestic	4.83	Low-emission double-glazing - domestic	2.07
Floor	Un-insulated solid ground floor	0.71	Un-insulated solid ground floor	0.71
Door	Wooden door	2.19	Wooden door	2.19

As a result, the total digital twin city energy consumption was reported to be 928 GWh and the total building energy intensity of 253 kWh/m^2. The carbon emissions for the city were also recorded, with 479123 tCO_2e for electrical emissions. Based on this data, the study found out that Bertam city had a value of 1.75 tonnes of carbon per capita. The building energy breakdown of the digital twin city also reported that most energy demands were for cooling purposes throughout the year. Scenario 2 – lowering U-value of building envelopes of Bertam city, was simulated after the baseline data was collected. The new total digital twin city energy consumption was recorded as 848 GWh. The carbon emissions rate also dropped to 434075 tCO_2e post optimization. Thus, iCD estimated energy savings and carbon reduction of 8.6% for Bertam city after the optimization works.

In this study, the master plan was divided into four zones of the sector, residential, commercial and institutional. Table 2 represented the data of energy consumption and carbon emissions collected from the simulation of both Scenario 1 and 2 in zoning sectors. The residential sector of Bertam, covering an area of 2.067 km^2 in Bertam city, was analyzed to produce a total of 2908 kWh, thus becoming the largest energy contributor to the overall energy consumption value of the city.

Table 2. Analysis of energy consumption and carbon emissions for Bertam city sustainable master plan of Scenario 1 and 2 based on zoning.

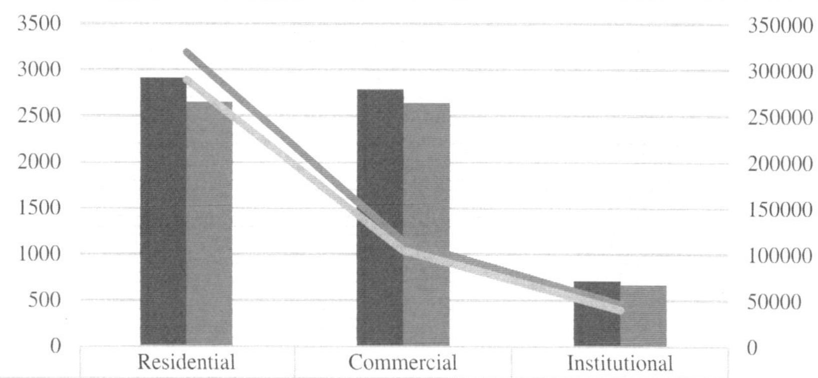

	Residential	Commercial	Institutional
Energy consumption baseline/kWh	2908	2781	712
Energy consumption optimized/kWh	2651	2637	668
Carbon emissions baseline/tCO2e	318929	113652.9	45760
Carbon reduction after optimization/tCO2e	289075	104076.7	39814.8

3.2. Heating, Ventilation & Air-Conditioning (HVAC) optimization on Bertam digital twin city

From the digital twin city simulation, this study has gained an annual building energy breakdown. Table 3 showed in detail the energy breakdown usage, separated into five uses, derived hot water, cooling, auxiliary, equipment and lighting. Cooling usage was noted as the highest energy consumer in Bertam city, conquering 36% of the total energy consumption, refer to table 3. High thermal exchanges between the inside and the outside of the buildings caused the inhabitants to depend on the HVAC system to achieve the desired thermal comfort. This ultimately caused a rising in electrical power demand for the buildings. Therefore, the study opted to convert the current HVAC system to another system with better energy performance in order to reduce the electrical load - Scenario 3. A comparison of building performance between Scenario 2 and 3 will be done to obtain the energy savings data. Variable Air Volume (VAV) indoor packaged cabinet system and fan coil system was used as the HVAC system in Scenario 2 and 3 respectively. A comparison analysis was made to calculate the potential energy savings and carbon reduction gained by changing to the better energy performance of the HVAC system, refer to Table 4. From the data shown, the study concluded an estimated potential savings and carbon reduction of 15.7% from the building operation simulation with fan coil system.

Table 3. Building energy breakdown analysis of Bertam in the year 2021.

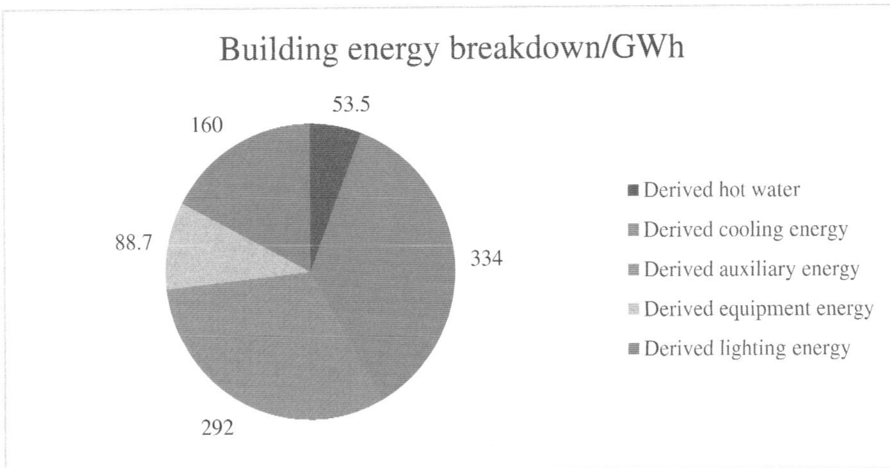

Table 4. Simulation result of estimated building energy performance between two different HVAC systems, Scenario 2 and 3 of the digital twin city.

Type of HVAC system	Total energy consumption/GWh
VAV Indoor Packaged Cabinet	848
Fan Coil System	715

3.3. Potential PV contribution for Bertam digital twin city

The use of renewable energy especially solar energy is inevitable for a tropical country such as Malaysia. Based on data from Petinrin, J.O., Bertam has an annual sun radiation of 1809 kWh/m [15]. Although it is not an uncommon value for a tropical country, Bertam still ranked among the highest in receiving solar radiation for cities in Malaysia. The iCD simulation is capable to calculate solar roof potential for the digital twin city model. Figure 2 showed a residential zone with an area of 0.324 km^2. The residential area was simulated in an early simulation to obtain the potential solar roof potential value – Scenario 4. The residential area was chosen as a precedent area because it is the largest residential zone with the highest data for building energy use intensity. The estimated solar roof potential for this area would be use as initial estimation PV contribution for the digital twin city. An analysis of total PV contribution and potential carbon emissions offset by PV panels was shown in Table 5. The total PV energy from the simulated area was estimated to be 14.7 GWh, out of 38 GWh of electrical energy for the residential. Meanwhile, the site yielded 220 kWh/m^2 of energy from the solar PV simulation. From table 5, this study has calculated a potential carbon emission offset and PV contribution of 21.95%.

Figure 2. The chosen residential area for potential roof solar PV energy performance simulation in Scenario 4.

Table 5. Potential solar PV energy performance analysis for the chosen residential area in Bertam city.

	Potential PV area/ km²	**PV yield energy/ GWh**	**Site PV yield energy intensity /kWh/m²**	**Total electrical energy/ GWh**	**Carbon emission from electricity/ tCO_2e**	**Emissions offset by PV/tCO_2e**
Residential area	0.066	14.7	220	38	19698	7623

4. Conclusion

This study presented a digital twin city model for sustainable city planning in Malaysia. Four different scenarios of a sustainable master plan occurred in the Bertam digital twin city to obtain an analysis of energy savings and carbon reduction. A comparison between Scenario 1 and Scenario 2, Scenario 2 and Scenario 3, and Scenario 3 and Scenario 4 showed energy savings of 8.6%, 15.7% and 24.95% respectively. The simulation of digital twin Bertam city has accumulated a total of 49.25% potential energy savings and carbon reduction post-optimization. NLCCF program target of carbon reduction of at least 33% by 2030 for Bertam city is achieved based on the analysis obtained from this simulation study. This digital twin city performance does not take into account the sustainable planning of a city for transportation, infrastructure and roads but only focuses on the building aspect. Therefore, the prospect of aiming zero-carbon city by 2050 is possible with absolute commitments between the town council, researchers and also the communities.

5. Acknowledgement

The work was supported by Universiti Kebangsaan Malaysia (UKM) DPK-2019-001; radiative heat transfer analysis of building envelope using Computational Fluid Dynamics.

References

[1] United Nations 2018 *Population Div. World Urbanization Prospects: 2018 revision* onl. Dept. Economics & Soc. Affairs

[2] Rasoolimanesh S M, Badarulzaman N, Abdullah A and Behrang M 2022 Integrated sustainable urban planning: a new agenda for future urban planning in Malaysia *J. Place Man. Dev.* **15** 3 284-97 https://doi.org/10.1108/JPMD-02-2020-0014

[3] Knoema 2019 Malaysia primary energy consumption https://knoema.com/atlas/Malaysia/Primary-energy-consumption ret. 28 Oct. 22

[4] Nasha L 2019 Low Carbon Cities – Malaysia's Response to Global Climate Emergency United Nations Development Programme https://www.undp.org/malaysia/news/low-carbon-cities-malaysias-response-global-climate-emergency ret 13 Dec. 19

[5] Rong T, Zhang P, Zhu H, Jiang L, Li Y, Liu Z 2022 Spatial correlation evolution and prediction scenario of land use carbon emissions in China *Eco. Informatics* **71** 101802 https://doi.org/10.1016/j.ecoinf.2022.101802

[6] Knoema 2022 Thailand CO_2 emissions per capita https://knoema.com/atlas/Thailand/CO2-emissions-per-capita

[7] Knoema 2022 Indonesia CO_2 emissions per capita https://knoema.com/atlas/Indonesia/CO2-emissions-per-capita

[8] Knoema 2022 India CO_2 emissions per capita https://knoema.com/atlas/India/CO2-emissions-per-capita

[9] Rosnazri A, Ismail D, Soib T 2012 A review on existing and future energy sources for electrical power generation in Malaysia *Ren. & Sust. Ener. Rev.* **16** 6 4047-55 https://doi.org/10.1016/j.rser.2012.03.003

[10] Suruhanjaya Tenaga 2020 Malaysia energy statistics handbook final energy consumption by sector p. 48 ST(P)03/03/2021 https://meih.st.gov.my/

[11] NLCCM 2021 *National low carbon cities masterplan* Min. of Env. & Water Malaysia vol 1 p 18

[12] Michael G 2016 Origins of the digital twin concept Florida Inst. Of Tech. & NASA 10.13140/RG.2.2.26367.61609

[13] Trauer J, Schweigert-Recksiek S, Engel C, Spreitzer K and Markus Z 2020 What is a digital twin? definitions and insights from an industrial case study in technical product development *Proc. design society: design conf.* vol 1 757-766 10.1017/dsd.2020.15

[14] Abdelali A, Mohamed T, Badr C, Emmanuel S, Abbas D and Karim A 2021 Towards a digital twin model for building energy management: case of Morocco *Procedia Comp. Sc.* **184** 401-10. https://doi.org/10/1016/j.procs.2021.03.051

[15] Petinrin, J O 2015 Renewable energy for continuous energy sustainability in Malaysia *Ren. Sust. Ener. Rev* **50** 967-81 10.1016/j.rser.2015.04.146

IOP Conf. Series: Materials Science and Engineering 1278 (2023) 012013 doi:10.1088/1757-899X/1278/1/012013

Thermal performance of reflective insulation system and comparison against ASHRAE Standard

M Z M Ashhar and C H Lim

Solar Energy Research Institute, Universiti Kebangsaan Malaysia, 43650 Bangi, Selangor Malaysia

chinhaw.lim@ukm.edu.my

Abstract. The trend in building energy consumption has been steadily increasing and will continue to rise significantly in the future. The application of a reflective insulation system in roof assembly is proven to be effective in reducing heat gain across the roof and can enhance indoor thermal comfort. This paper presents the development of an Indoor Solar Simulator to evaluate the thermal resistance (RSI value) of reflective insulation systems for roof assemblies. This paper also compares RSI values obtained experimentally and the RSI values listed in ASHRAE Standard. The roof assemblies studied in this paper incorporate the same reflective insulation systems as listed in ASHRAE Standard. Thermocouples and heat flux transducers are installed at various locations and the RSI values are calculated using the average method as detailed in ISO 9869-1:2014. From the error analysis, it can be concluded the RSI values obtained experimentally using the Indoor Solar Simulator and ASHRAE Standard are in a good agreement with MAPE value of 6.13%. These discrepancies could be due to the thermal bridges from mechanical supports inside the air gap, or possibly the air gaps are not completely airtight and unventilated.

1. Introduction

In 2020, the building sector accounted for 36% of global final energy consumption and 37% of energy-related carbon emissions. Although carbon emissions dropped slightly compared to the previous year due to the global pandemic, the reopening of the economic sector is expected to increase the emissions. One of the major contributions to high carbon emissions and high energy consumption in buildings is space cooling. It is responsible for 5% of total energy consumption worldwide in 2020. Furthermore, the energy use for space cooling is expected to double in 2040 [1]. This poses a huge challenge, especially for countries in the Tropics region where cooling is dominant as air-conditioning can make up to 50% of the total electricity usage [2].

All year long, Malaysia experiences hot and humid weather, which causes a tremendous amount of heat gain in buildings. High energy demand for cooling has resulted from the choice of construction materials, particularly the roof [3]. A significant portion of the solar energy entering a building is absorbed by exposed roof surfaces and transferred into the building. When compared to other building elements, the roof is the one that causes the greatest temperature variations in single-story homes or low-rise buildings [4]. According to Nahar et al. [5], the roof contributes 36.7% of the heat load in a single-story home. According to Vijaykumar et al. [6], between 50% and 70% of heat is transmitted through the roofs of single-story or double-story buildings. Conduction, convection, and radiation are how solar heat enters a structure. Radiation accounts for about 87% of the total heat energy that enters the building through the roof, with conduction and convection heat transfer providing the balance [7]. Concrete tiles, clay tiles, and metal decks are a few of the often-used roofing materials in Malaysian homes. These roofing materials allow a lot of solar heat into buildings, raising interior temperatures, using more energy, and making indoor spaces uncomfortable [8].

1.1. Reflective insulation system

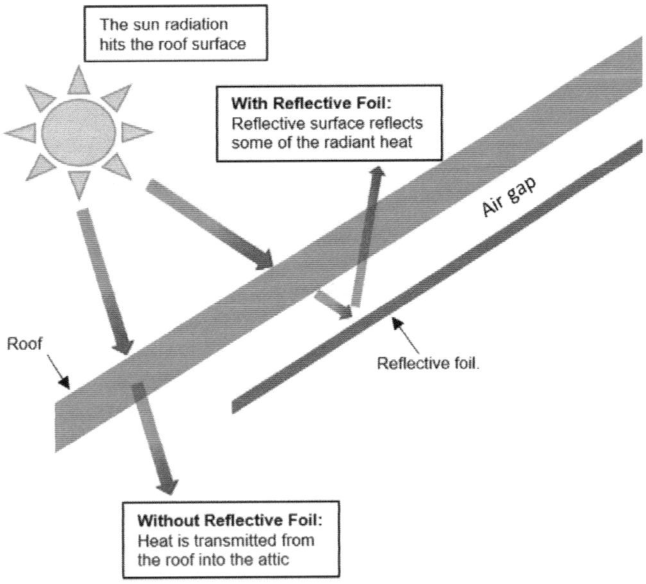

Figure 1. Working principle of reflective insulation system in a roof system [10]

According to Reflective Insulation Manufacturers Association International or RIMA-I, a reflective insulation system is when a reflective foil with a low emissivity characteristic on one or both sides of the foil, is installed adjacent to an air gap. The low emissivity side of the foil ($\varepsilon = 0.01 - 0.05$) must be installed adjacent to an air gap for the system to reflect radiant heat transfer effectively as shown in Figure 1. However, the opposite effect is observed if the reflective foil is stacked between two solid materials, as the conductive heat transfer will be high. Based on the previous literature, there are various factors affecting the performance of a reflective insulation system as a roof insulation, such as the emissivity value, thickness of the air gap, air gap ventilation, roof pitch, dusting effect, weather, and climatic conditions [10].

1.2. Previous work

The use of an indoor solar simulator to investigate roof thermal performance has been utilised by previous researchers. Researchers favoured this method as it allows the research to be conducted in a steady-state environment and the modifications to the roof assembly can be done immediately. It was proven that an indoor solar simulator can produce a reliable and accurate measurement in evaluating roof thermal performance.

There were some similarities and differences between the previous indoor solar simulators. In terms of the choice of the light source, Chang et al. [11] used halogen light bulbs, whereas Roels & Deurinck [12] used infrared light bulbs to generate solar irradiance of 600 W/m². In contrast, Shrestha et al. [13] heated the roof surface and maintained it at 60°C using infrared light bulbs to simulate summer conditions. One significant benefit of using an indoor solar simulator is the ease of changing the roof configurations and test parameters. For example, Roels & Deurinck [12] and Shrestha et al. [13] were able to change the type of radiant barrier, whereas Chang et al. [11] were able to change the position of the radiant barrier. Most importantly, steady-state measurements can be obtained as the research is conducted in a controlled indoor environment which is vital to obtain consistent results and not influenced by the transient outdoor environment. Although previous research has shown that steady-state measurements can be obtained via indoor laboratory experiments, none of the past research has made a comparison or validation against ASHRAE Standard.

1.3 RSI values of enclosed air gaps as listed in the ASHRAE Standard

Previously, The American Society of Heating, Refrigerating, and Air-Conditioning Engineers (ASHRAE) has provided the RSI values of reflective insulation systems (enclosed air gaps facing a low emissivity surface). The RSI values of enclosed reflective air gaps are based on the reduced radiative transport across the air gaps and convection/conduction occurring in the air gap, so air film resistances are included. Table 2 below shows the RSI values of enclosed air gaps facing surfaces of different emissivity (ε = 0.03, 0.05, 0.2, 0.5, 0.82) and air gap thicknesses from 12.7 mm to 139.7 mm. The angle of the enclosed air gap is kept at 0° (horizontal plane), and the heat direction is downwards from top to bottom surface.

Table 1. The RSI values of enclosed air gaps listed in the ASHRAE Standard [14]

	RSI value (m^2K/W)				
Air gap thickness (mm)	ε = 0.03	ε = 0.05	ε = 0.2	ε = 0.5	ε = 0.82
12.7	0.437	0.412	0.294	0.187	0.136
19	0.625	0.579	0.370	0.215	0.150
38	1.073	0.942	0.491	0.252	0.166
88.9	1.774	1.442	0.601	0.277	0.176
139.7	2.064	1.627	0.631	0.284	0.178

1.4 Research objective

This paper presents the development of an indoor solar simulator to investigate the thermal resistance (RSI values) of reflective insulation systems for roof applications. This research focuses on validating the Indoor Solar Simulator test results and comparison with RSI values of enclosed air gaps presented in the ASHRAE Standard.

2. Details of the indoor solar simulator for the roof thermal resistance measurement

For this research, an indoor solar simulator was built in Universiti Kebangsaan Malaysia in Selangor, Malaysia (latitude 2°55'16.4"N and longitude 101°46'15.6"E). According to the Koppen-Geiger climate classification, the climate in Malaysia is categorised as Tropical Rainforest. Malaysia has a maximum temperature between 29°C and 30.3°C and high relative humidity between 82% and 90% annually. Besides that, Malaysia also suffers from a lot of rain even during the driest month with an annual rainfall of 2981 mm [15].

(a) (b)

Figure 2. (a) Halogen lamps used to roof surface (b) The overall view of the indoor solar simulator

The indoor solar simulator aims to evaluate the RSI value of various types of roof configurations under hot and humid conditions. The indoor solar simulator was designed in such a way that the roof can be assembled and disassembled easily so that various designs of roof configurations can be installed and measured conveniently. The indoor solar simulator is exposed to a steady-state indoor environment with an average indoor temperature of 31°C and 80% relative humidity. The wind effect is assumed to be negligible as there is minimal air movement inside the building. The steady-state indoor conditions ensure more accurate and repeatable measurements, with a shorter time taken to complete each measurement.

2.1. Sensors and data acquisition system

To analyse the thermal response of the roof assembly, three pieces of thermocouples are positioned on the outer roof surface (T_{1A}, T_{1B}, T_{1C}) and three pieces of thermocouples are positioned on the inner roof surface (T_{2A}, T_{2B}, and T_{2C}). Two pieces of heat flux transducers are placed on the inner roof surface (q_1 and q_2). The signals transmitted by the sensors are logged at the one-minute interval by the Arduino controller and data acquisition system. Each sensor has been calibrated within the previous 24 months. The roof inclination angle can be varied from 0° (horizontal) to 90° (vertical). The test specimen size is 1.5 m in width and 3 m in length. Figure 3 shows the test setup and the positions of the sensors.

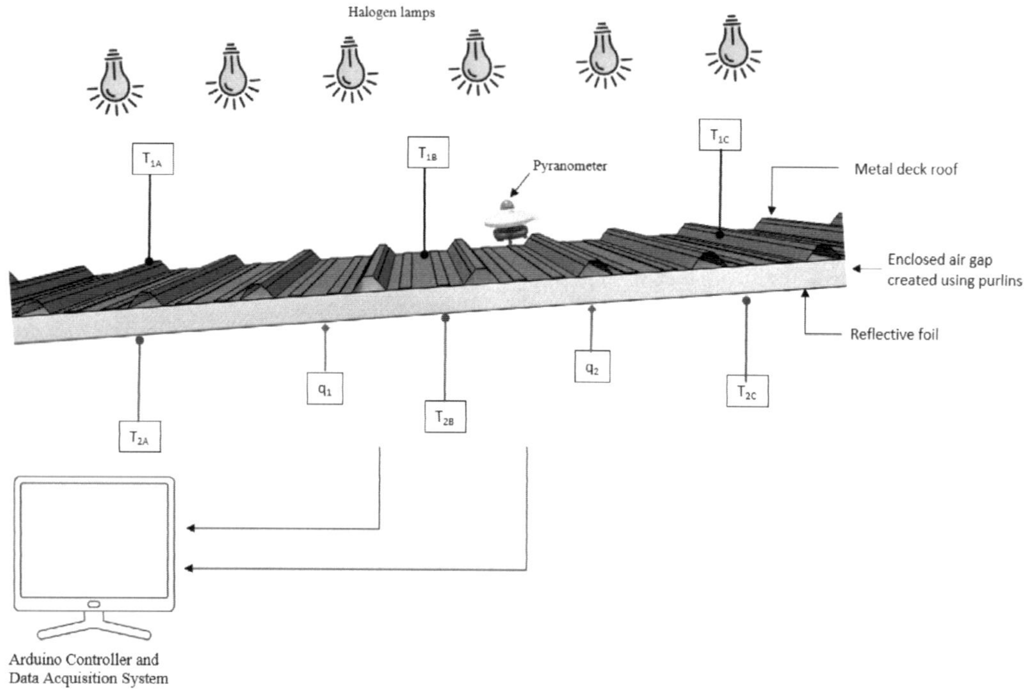

Figure 3. Positions of the thermocouples and heat flux transducers on the test assembly

3. Methodology

3.1. Validation of roof surface temperature against outdoor measurement

To ensure a realistic experiment is conducted, the indoor solar simulator measurement of roof surface temperature was first validated against outdoor measurement. The outdoor heat flux measurement is nearly impossible to obtain due to the unsteady-state condition of the outdoor environment. The roof assembly used for validation is a Metal deck roof with polyurethane (PU) insulation.

Figure 4. Measuring the roof surface temperature of a metal deck roof on a hot sunny day in Malaysia

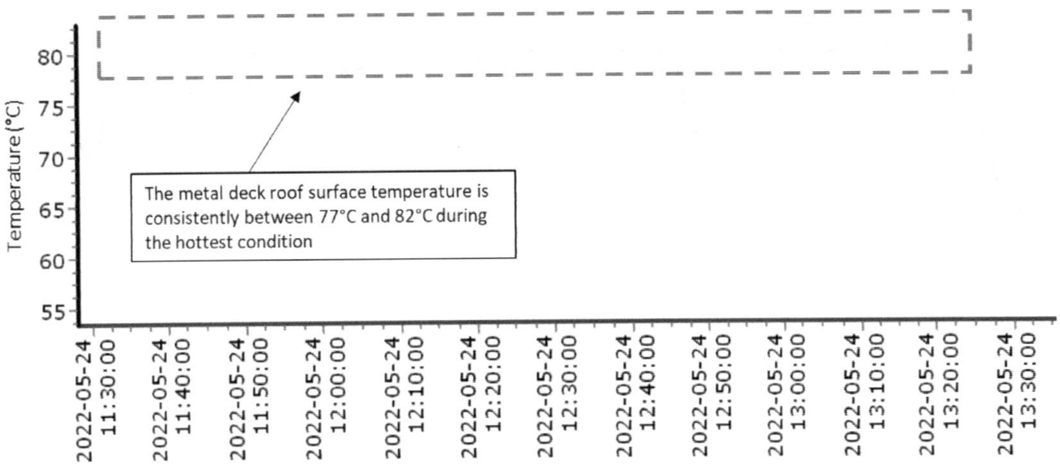

Figure 5. Roof surface temperature of metal deck roof at an outdoor condition during the hottest condition of a sunny day in Malaysia

The outdoor surface temperature of the metal deck roof with PU insulation was taken during the daytime. It was found that the maximum outdoor surface temperature reaches a maximum constant temperature of 77°C – 82°C in the afternoon. For the indoor solar simulator measurement, the power supplied to the halogen lamps is adjusted so that the outer roof surface temperature of the metal deck roof with PU insulation reaches a constant temperature of 77°C – 82°C, mirroring the values obtained from the outdoor measurement. The same amount of power supply is applied for all measurements using the indoor solar simulator.

3.2. Data collection

The desired roof configuration is assembled on the indoor solar simulator. After the halogen lamps are turned on, the roof surface temperature and heat flux readings are constantly monitored and recorded every 2 seconds. The readings typically reach a steady-state condition 1 hour after the lamps are turned on. The measurement is continued for an extra 3 hours after the steady-state condition was achieved to ensure a stable and accurate reading is obtained. The roof surface temperatures (T_{1A}, T_{1B}, T_{1C}, T_{2A}, T_{2B}, and T_{2C}), and heat flux transmissions (q_1 and q_2) in the steady-state period are extracted to calculate the RSI value of the entire roof assembly.

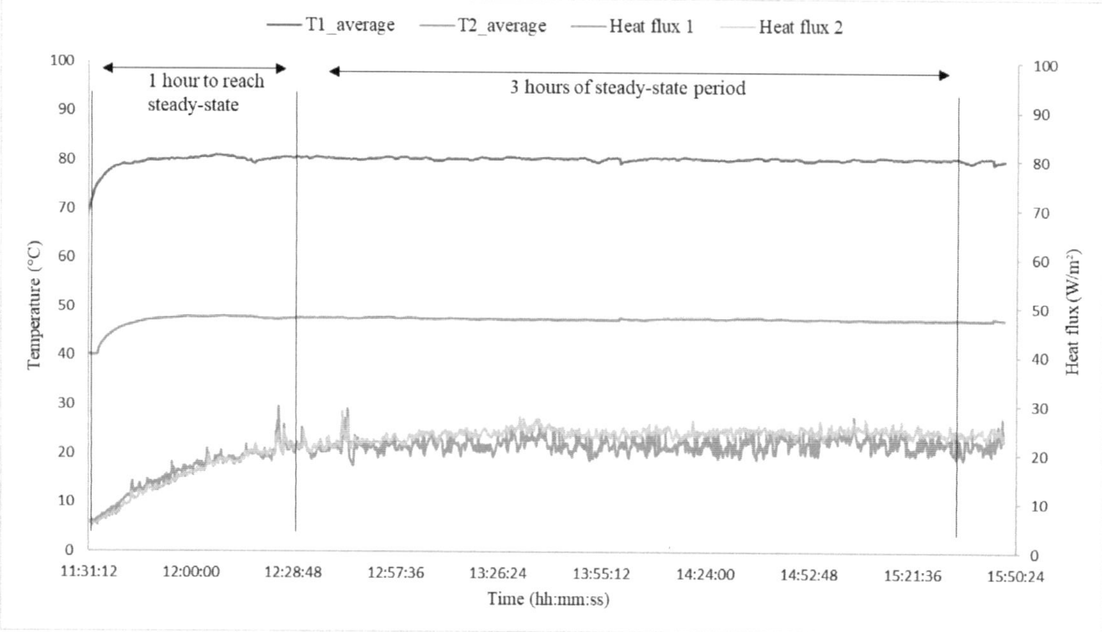

Figure 6. Measurement of roof surface temperature and heat flux penetration using the indoor solar simulator. A Steady state is achieved 1 hour after measurement starts.

3.3. *Thermal Resistance (RSI value) calculation method*

The method chosen to determine the RSI value of the entire roof assembly is the average method as detailed in ISO 9869-1:2014. The calculation is achieved by summing up the average temperature differences between the outer roof surface and inner roof surface (T_{1_avg}-T_{2_avg}) and dividing them by the sum of all averaged heat fluxes ($q_{\_avg}$). The surface temperature and heat flux values used for the RSI value calculation are the values in the steady-state period. This step is repeated for all roof configurations studied in this research.

$$RSI = \frac{\sum(T_{1\_avg} - T_{2\_avg})}{\sum(q_{avg})} \quad (1)$$

where:
T_{1_avg} = Average values of T_{1A}, T_{1B}, and T_{1C}
T_{2_avg} = Average values of T_{2A}, T_{2B}, and T_{2C}
$q_{\_avg}$ = Average values of q_1 and q_2

3.4. *Specification of the roof configurations in this study*

The roof configurations in this study are aimed to mimic the assembly as reported by ASHRAE Standard. Therefore, the reflective insulation systems are made up of metal deck shingles with a reflective foil (ε = 0.03) separated by an enclosed air gap created using purlins. The reflective foil's low emissivity side faces the enclosed air gap (upwards) for all roof configurations. The thicknesses of enclosed air gaps are 25 mm, 50 mm, 75 mm, 100 mm, and 125 mm. The roof pitch of all roof configurations is maintained at 0° (horizontal).

IOP Conf. Series: Materials Science and Engineering 1278 (2023) 012013 doi:10.1088/1757-899X/1278/1/012013

Table 2. Illustration and description of the studied roof configurations

Number	Illustration of the roof configuration	Description
1	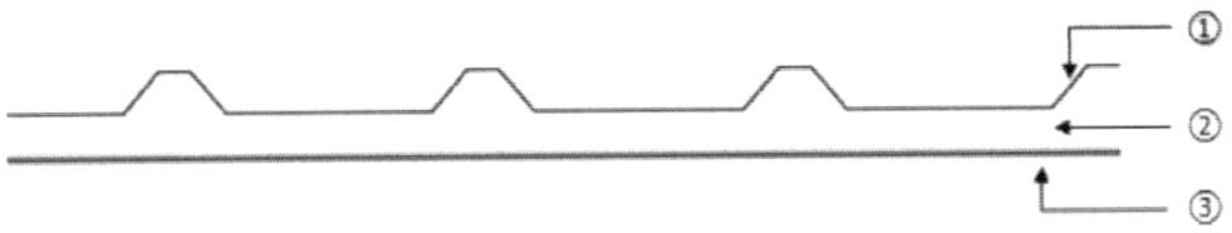	1. Metal deck roof 2. Air gap 25 mm 3. Reflective foil (ε = 0.03)
2	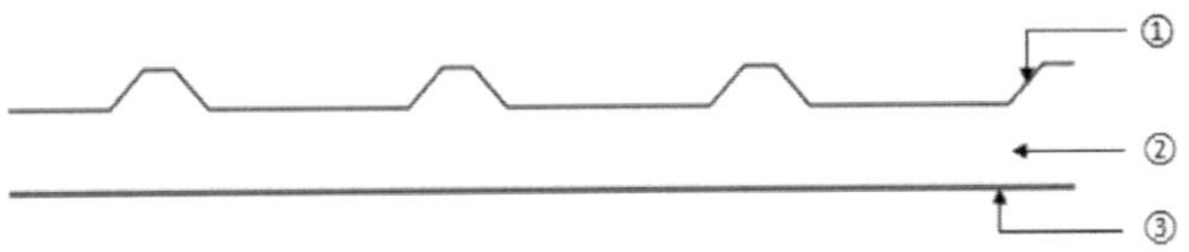	1. Metal deck roof 2. Air gap 50 mm 3. Reflective foil (ε = 0.03)
3	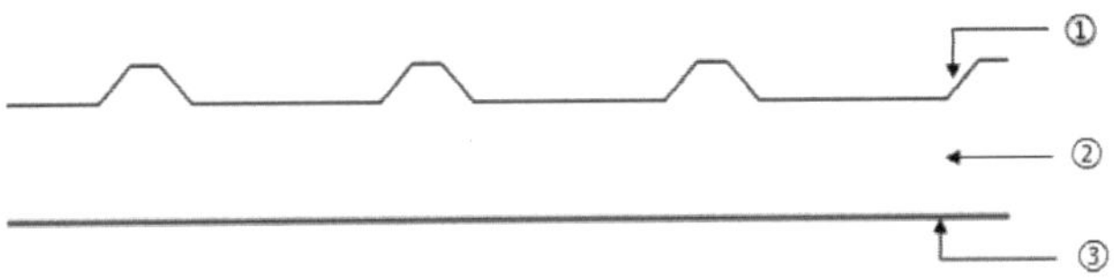	1. Metal deck roof 2. Air gap 75 mm 3. Reflective foil (ε = 0.03)
4	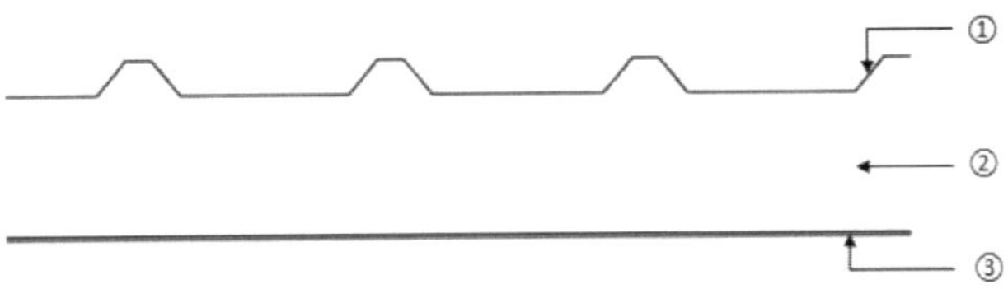	1. Metal deck roof 2. Air gap 100 mm 3. Reflective foil (ε = 0.03)
5	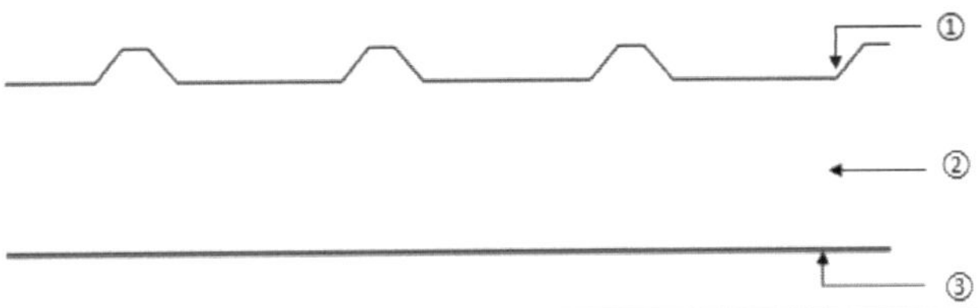	1. Metal deck roof 2. Air gap 125 mm 3. Reflective foil (ε = 0.03)

4. Results and Discussions

Figure 8 below shows the RSI values of roof assembly obtained experimentally against ASHRAE Standard for various air gap thicknesses. The ASHRAE Standard values are shown in Table 2 (ε = 0.03). It can be seen that the RSI values obtained experimentally are always higher than ASHRAE Standard for all air gap thicknesses despite both plots having similar trends.

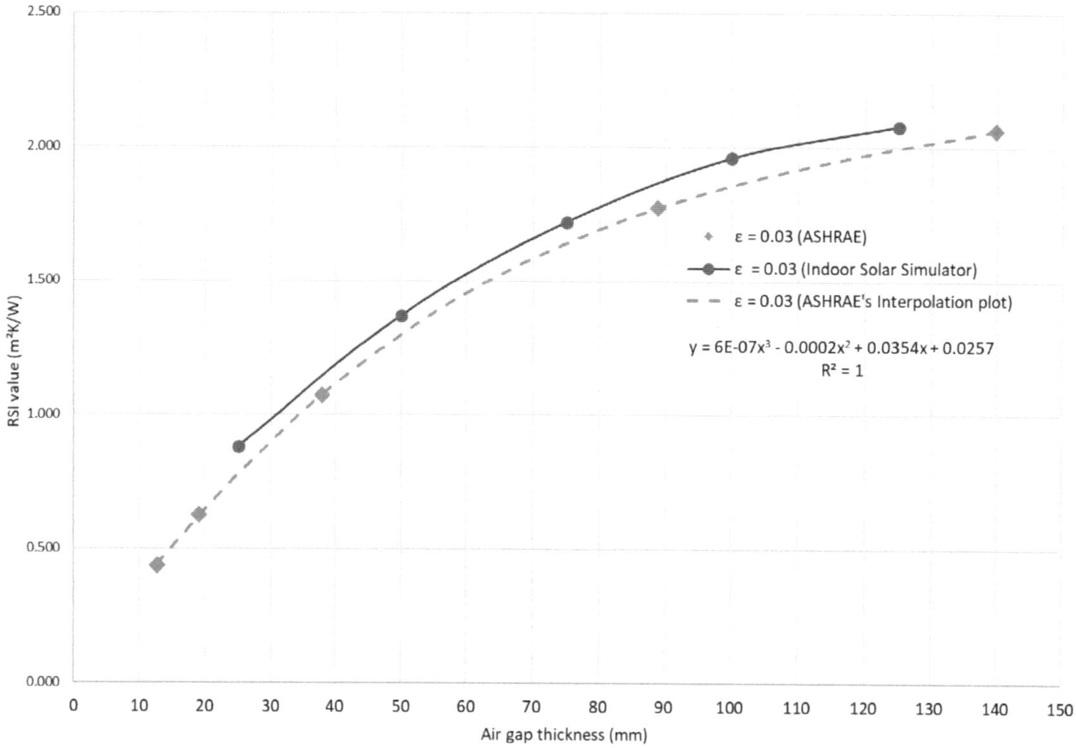

Figure 7. RSI values of roof assembly obtained from the Indoor Solar Simulator and ASHRAE Standard

Table 3. The percentage error of RSI values obtained from the Indoor Solar Simulator and ASHRAE Standard

Air gap thickness (mm)	RSI value from ASHRAE's Interpolation (m^2K/W)	RSI value from Indoor Solar Simulator(m^2K/W)	Percentage error (%)
25	0.78	0.88	11.60
50	1.30	1.37	5.17
75	1.64	1.72	4.58
100	1.86	1.96	5.29
125	2.00	2.08	4.03
Mean absolute percentage error (MAPE)			**6.13**

Table 7 shows the percentage error of RSI values between the Indoor Solar Simulator and the ASHRAE Standard. As the air gap thicknesses listed in ASHRAE Standard (as shown in Table 2) and the air gaps evaluated using Indoor Solar Simulator are not the same, hence an interpolation plot was used to estimate the ASHRAE's RSI values roof assemblies with air gap thickness of 25 mm, 50 mm,

75 mm, 100 mm, and 125 mm. The percentage error is about 4% - 5% for air gap thickness between 50 mm and 125 mm. For the air gap of 25 mm, however, the percentage is quite high at 11.60%. From the error analysis, it can be concluded the RSI values obtained experimentally using the Indoor Solar Simulator and ASHRAE Standard are in a good agreement with MAPE value of 6.13%.
The RSI values in ASHRAE Standard are obtained for ideal conditions with simplified assumptions such as disregarding any thermal bridges from mechanical supports inside the air gap, and the air gaps are completely airtight and unventilated [16]. These conditions are almost impossible to achieve in an experimental condition and which justifies the errors obtained experimentally using the Indoor Solar Simulator.

5. Conclusion

The paper presents the development of an indoor solar simulator for evaluating the roof's thermal performance in terms of RSI value and comparison against the RSI values listed in the ASHRAE Standard. Halogen light bulbs are used to simulate sun radiation which is directed toward the outer roof surface. Thermocouples are installed on the outer and inner roof surfaces, and heat flux sensors are installed on the inner surface. The surface temperature of various roof surfaces and heat flux transmissions are recorded 3 hours after steady-state conditions are achieved. RSI values are calculated using the average method as detailed in ISO 9869-1:2014. The error analysis showed that the RSI values obtained experimentally and in ASHRAE Standard are in good agreement with MAPE value of 6.13%. Several reasons for the errors are because of thermal bridges in the air gaps, which occur in the experimental roof assembly, and lack of airtightness which causes the air inside the air gap to be ventilated. As the workmanship of the roof assembly can affect the accuracy of the measurements, the researchers must ensure the roofs are assembled properly, and the air gaps are always airtight and unventilated.

6. References

[1] United Nations, "Global Status report for Buildings and Construction 2021," *United Nations Environ. Program.*, 2021, [Online]. Available: https://globalabc.org/resources/publications/2021-global-status-report-buildings-and-construction.
[2] International Energy Agency (IEA), "'The Future of Cooling Opportunities for energy-efficient air conditioning'International Energy Agency Website: www.iea.org, 2018," 2018, [Online]. Available: www.iea.org.
[3] A. M. Abdul Rahman, N. M. S. @ Abd.Rahim, K. Al-Obaidi, M. Ismail, and L. Yoke Mui, "Rethinking the Malaysian Affordable Housing Design Typology in View of Global Warming Considerations," *J. Sustain. Dev.*, vol. 6, no. 7, pp. 134–146, 2013, doi: 10.5539/jsd.v6n7p134.
[4] I. Hernández-Pérez, G. Álvarez, J. Xamán, I. Zavala-Guillén, J. Arce, and E. Simá, "Thermal performance of reflective materials applied to exterior building components - A review," *Energy Build.*, vol. 80, pp. 81–105, 2014, doi: 10.1016/j.enbuild.2014.05.008.
[5] Nahar N, Sharma P, and P. M., "Performance of Different Passive Techniques for Cooling of Buildings in Arid Regions," *Build. Environ.*, vol. 38, no. 1, pp. 109–116, 2003, doi: doi:10.1016/S0360-1323(02)00029-X.
[6] K. C. K. Vijaykumar, P. S. S. Srinivasan, and S. Dhandapani, "A performance of hollow clay tile (HCT) laid reinforced cement concrete (RCC) roof for tropical summer climates," *Energy Build.*, vol. 39, no. 8, pp. 886–892, 2007, doi: 10.1016/j.enbuild.2006.05.009.
[7] F. Richards and D. Hassal, *Reflective Insulation and the Control of Thermal Environments*, Metric edi. Sydney, Australia, 1977.
[8] K. M. . Al-Obaidi, M. Ismail, and A. M. Abdul Rahman, "Passive cooling techniques through reflective and radiative roofs in tropical houses in Southeast Asia: A literature review," *Front.*

Archit. Res., vol. 3, no. 3, pp. 283–297, 2014, doi: 10.1016/j.foar.2014.06.002.

[9] Department of Standards Malaysia, *Energy efficiency and use of renewable energy for non-residential buildings - Code of practice (MS 1525:2014)*. 2014.

[10] M. Z. Mohd Ashhar and L. C. Haw, "Recent research and development on the use of reflective technology in buildings – A review," *J. Build. Eng.*, vol. 45, no. October 2021, p. 103552, 2021, doi: 10.1016/j.jobe.2021.103552.

[11] P. C. Chang, C. M. Chiang, and C. M. Lai, "Development and preliminary evaluation of double roof prototypes incorporating RBS (radiant barrier system)," *Energy Build.*, vol. 40, no. 2, pp. 140–147, 2008, doi: 10.1016/j.enbuild.2007.01.021.

[12] S. Roels and M. Deurinck, "The effect of a reflective underlay on the global thermal behaviour of pitched roofs," *Build. Environ.*, vol. 46, no. 1, pp. 134–143, 2011, doi: 10.1016/j.buildenv.2010.07.005.

[13] S. Shrestha, W. Miller, and A. Desjarlais, "Thermal Performance Evaluation of Attic Radiant Barrier Systems Using the Large Scale Climate Simulator (LSCS)," *ASHRAE Trans.*, 2013.

[14] ASHRAE, "Chap. 26: Heat, air and moisture control in building assemblies - Material properties," in *ASHRAE Handbook of Fundamentals*, 2009, pp. 26.1--26.22.

[15] Climate-Data, "Climate: KUALA LUMPUR," 2021. https://en.climate-data.org/asia/malaysia/kuala-lumpur/kuala-lumpur-715107/.

[16] A. Pourghorban and B. M. Kari, "Evaluation of reflective insulation systems in wall application by guarded hot box apparatus, and comparative investigation with ASHRAE and ISO 15099," *Constr. Build. Mater.*, vol. 207, pp. 84–97, 2019, doi: 10.1016/j.conbuildmat.2019.02.097.

IOP Conf. Series: Materials Science and Engineering 1278 (2023) 012014 doi:10.1088/1757-899X/1278/1/012014

Mesh sensitivity analysis for Proton Exchange Membrane Fuel Cells using ANSYS Fluent

Abshir M. Ashour[1*], Adnan Ibrahim[1], Kamaruzzaman Sopian[1], Khaled S. Alzahrani[2]

[1]Solar Energy Research Institute, University Kebangsaan Malaysia, 43600 Bangi, Selangor, Malaysia

[2]Department of Mechanical Engineering Technology, Yanbu Industrial College, 41912 Yanbu Alsinaiyah, Saudi Arabia

*Correspondence: amahmed1@asu.edu; Tel: +966-565-241-454; Fax:+966-143-946-144

Abstract: Improving the design of Proton Exchange Membrane Fuel Cells (PEMFC) is crucial but requires a deep understanding of the basic science surrounding the technology. In this study, the membrane electrolyte assembly (MEA), catalyst layers channels, and gas diffusion layers are modelled using the commercial PEMFC simulation package built into ANSYS FLUENT 12.1. With various mesh sizes of 20 ,40 and 80 cells, a mesh sensitivity study of a cross-section of the fuel cell membrane, anode, and cathode, was carried out while 16000 data iterations were collected. The simulation results showed that all data points of membrane and anode layers converged before or at 10,000 iterations, while varying cathode needs more time to converge. A study of iterations and convergence, resulting in the comparison of various meshes at different data points, is discussed while a comprehensive description of the simulation's parameters is provided, and the emphasis is placed on determining the convergence criteria that are necessary for delivering consistent results. One of the purposes of this experiment is to demonstrate that high-quality computer modelling and simulation of fuel cells can accelerate the discovery of fuel cell components that are optimally designed. As a result, the study confirmed that computer modelling can contribute to a better understanding the mechanics and reactions occurring within the fuel cell and its layers in order to improve its performance and lifecycle in the future.

Keywords: Proton Exchange Membrane Fuel Cell; Modelling and Simulation; Computational Fluid Dynamics; iteration; convergence.

1. Introduction

Proton exchange membrane fuel cells (PEMFCs) are considered one of the most promising clean sources of electrical energy. PEMFC, which converts chemical energy into electrical energy, can potentially replace the batteries and the internal combustion engine because they are clean, quiet, modular, and capable of a quick startup [1]. However, before this technology can be commercialized and widely used, various technical challenges need to be addressed. These include improvements in performance, reliability, durability, size, weight, availability of hydrogen fuel, and reductions in production costs [2]. One way to expedite its technological development is by using computer modeling [3]. Because computer Modeling assists in decreasing the cost and accelerating the speed at which designs can be evaluated while enhancing the fundamental understanding of the operation and performance of fuel cells. As computing power prices continue to decline, simulations may be done relatively fast and affordably. A study by Arvay et al. on PEMFC membrane and catalyst layer mesh sensitivity using ANSYS indicated that more than 10,000 iterations are required to get within 5% of the value at 19,000 iterations [4]. However, based on these findings, it appears that varying the number of computational cells within the membrane has little effect, while varying the catalyst layer does have a noticeable impact. This demonstrates that their solution does not converge because the results of 20 divisions revealed decreased current densities, whereas the overall current density increased until ten divisions. One of the recommendations of this study was to investigate and analyze the mesh sensitivity of Proton Exchange Membrane fuel cell components.

Iranzo et al.[5] developed a simulation tool for geometrical analysis and optimization of fuel cellbipolar plates. The focus of this model was to analyze the geometrical influence of bipolar plates in fuel cells. The model predicted electrochemistry, liquid water generation, mass continuity, current continuity, membrane humidification, energy evacuation, electric losses, and other phenomena. The findings of this research demonstrated that the model was able to generate accurate predictions regarding the behavior of fuel cells. Cheddie et al. [6] presented a three-dimensional computer modeling to understand the operation of a polybenzimidazole membrane fuel cell. This model originated from a previous two-dimensional and one-dimensional model study. The largest oxygen depletion occurs in the cathode catalyst layer directly beneath the ribs, according to the results. This depletion rises in the flow direction and is more evident at lower gas supply flow rates. Siegel et al. [7] developed modeling techniques and made key assumptions that are now widely employed. They also summarized the computational techniques and popular numerical algorithms used for PEMFC studies. It was demonstrated that one- and two-dimensional models are enough for a variety of forms of research. Diverse modeling methodologies, their implementations, as well as their benefits and drawbacks, were discussed. A study of the available PEM modeling tools revealed that Ansys Fluent was utilized in the majority of the published works. By employing a three-dimensional computer model, Kuo et al. [8] investigated the performance characteristics of PEMFC. This study emphasized the effect of the wave-like surface on the electrochemical reaction efficiency and the electrical performance of the PEMFC at various operating temperatures. Their results showed a dramatic increase in the catalytic reaction's efficiency. The enhanced catalytic reaction produces increased cell voltage and power density, especially at greater values of the inlet temperature and current density. Bednarek and Tsotridis performed a simulation study on a single channel fuel cell model. They highlight possible limitations and difficulties associated with CFD in PEMFC modeling. Due to inherent numerical instabilities, an appropriate convergence methodology should be applied for steady-state solutions. Their results showed that the lack of reliable and robust input data is one of the issues for obtaining accurate results [9].

In recent years, a comprehensive review of the state of the art of PEMFC can be found discussingthe modeling methods and the diagnostic validation techniques in PEM technology [10]. Pant et al. developed a modified binary friction model (MBFM), capable of predicting mass transport in viscousand transitional Knudsen regions [11]. The study suggested that the choice of transport model can significantly impact the predictions of mass transport. The predictions of MBFM are significantly different compared to the the existing models for flow in the Knudsen region and mixtures containinggases with a large difference in molecular weights. Gößling et al. performed a simulation-based investigation of the operation of a PEMFC system focusing on its water management by developing a2-D + 1-D PEM fuel cell stack model, which was an integrated AVL Cruise-M fuel cell system simulation. An analysis is presented comparing a system operation with a fuel cell in co- and counter-flow configuration with a special focus on the local and overall water management [12]. A comprehensive CFD-based tool to model is developed using Open FOAM in work presented by Kone et al. [13]. It can be used to rapidly gain insights into the cell working processes, whereby the base case I-V curve results showed better agreement with the experimental trend at low current densities. The simulation indicates that the chosen concentration constant significantly impacts the overpotential concentration. Wilberforce et al. investigated the effects of material (open-pore cellular foam material)and structural geometry (serpentine) of a PEMFC using ANSYS to measure the pressure drop across the structure [14].

Techniques such as computational modeling, simulation, and experimental are also employed tostudy the behavior and performance of the PEMFC membrane. An empirical diffusion model for Nafion, designed exclusively for Nafion by Ren et al. [15], is the foundation for several membrane models. There have been several studies into the formation of water and flooding inflow channels, including [16] and [17]. Shah et al. [18] provided a comprehensive review that focused on advances in degradation and microscale phenomena modeling, which was published in 2011. An experimental technique is used by [19] to investigate the impact of the hydrogen circulation rate, which is a critical parameter of the hydrogen circulation anode mode, on water management and cell performance. Xiangchao et al. [20] have simulated and analyzed the effect of operating

IOP Conf. Series: Materials Science and Engineering 1278 (2023) 012014 doi:10.1088/1757-899X/1278/1/012014

variables on the performanceof a PEMFC with anode recirculation. These variables include the cathode inlet relative humidity, cathode and anode intake pressure, and anode stoichiometry. Based on the gaps derived from previous investigations, this study was performed with the aim to improve the computer modeling capability and accuracy in the design and analysis of PEMFC. Whilemesh sensitivity analyses have been performed in the previous works, they are focused mainly on one parameter. This work extended the scope by combining multiple cases – mesh layers, membrane, anode mesh, and cathode mesh. An ANSYS computational model and simulation were used in this work to perform mesh sensitivity analyses in the interior structures of PEMFC.

2. The PEM fuel cell operating principle and its performance

The basic structure of a PEM fuel cell, its crucial components, and the general operating principlesare depicted in Figure 1. These components will include electrolytes, electrodes, interconnects, the current collector, and catalyst material [21]. The electrolyte in this fuel cell is a complex polymer, often referred to as a "membrane", and the membrane is an excellent proton conductor. Typically, a thin organic polymer (poly-perfluoro sulfonic acid) called Nafion is used as an electrolyte membranefor the PEMFC [22]. This polymer electrolyte membrane can produce high-power density with the advantage of having a low weight and low volume. The carbon electrodes with platinum electro- catalyst or platinum-ruthenium alloy are used for both the anode and cathode sides and with carbon or high-qualitymetal interconnects [23].

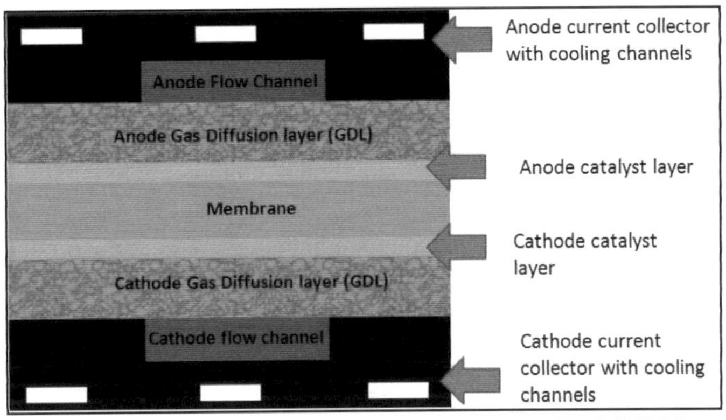

Figure 1. Schematic of a PEMFC from the Ansys Fuel Cell Manual.

Oxygen passes over one electrode and hydrogen passes over the other, generating electricity, water, and heat. In other words, PEM fuel cells convert hydrogen and oxygen to water and produce electricity and heat. Figure 2 shows the overall electrochemical energy conversion process in PEMFC.This illustrates that the PEM fuel cells run on pure hydrogen and produce zero emissions at the point of use, creating the potential for a clean, sustainable energy cycle. Unlike the other fuel cells, the PEM fuel cells need only hydrogen, water, and oxygen from the air to operate, while other fuel cells need corrosive fluids. PEM fuel cells demonstrated that they can produce current densities of over 2 kW/l and 2 W/cm^2. The PEMFC operates at less than 100°C, and its efficiency can reach up to 60% if pure H_2 is used as fuel [24]. These PEM fuel cells can be used as electrochemical engines for stationary and automotive applications. The following are the chemical reactions that occur in a PEM fuel cell [25]:

H_2 e 2H+ + 2e- (anode reaction) (1)
½ O_2 + 2e- + 2H+ eH_2O (cathode reaction) (2)

The following describes the entire process that takes place in a PEMFC fuel cell.

$2H_2$ + O_2 e $2H_2O$ + Electricity + Heat (3)

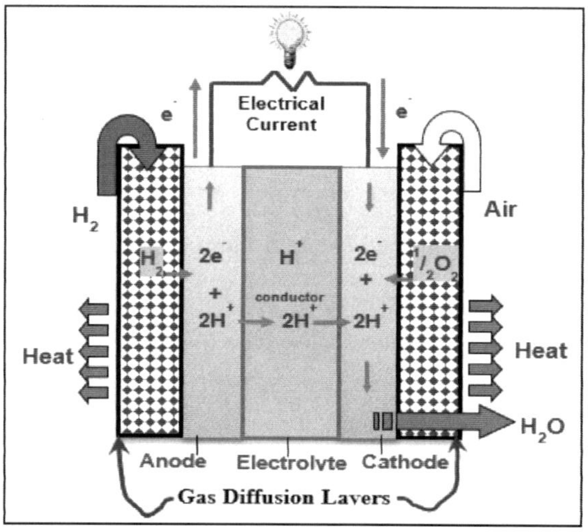

Figure 2. Schematic and operation illustration of a basic PEM fuel cell.

3. Methodology

The mesh sensitivity analysis was carried out using a three-dimensional (3D) computational fluid dynamics model of a single cell proton exchange membrane fuel cell (PEMFC) [26] and [27]. The initial step was to generate the fuel cell geometry using computer-aided design tools, in this case, Solid-Works, which was used to create the geometry for this project. This geometry was created by using a single channel serpentine design model. This design incorporates a single channel that constantly winds in a single path from the inlet to the outflow. The second phase involves decomposing the geometry into a computational mesh and determining its boundary conditions. The Cartesian grid approach was used to create a mesh for the model. Finally, the simulation's physical and operating parameters are entered into the simulation during the third and final steps. The thermal and electrical properties of various materials and operating temperatures, inlet gas flow rates, pressures, open-circuitvoltage, humidification, and porosity are just a few of the variables that were considered.

The geometrical model serves as the foundation for the construction of a computational mesh. Among the wide range of mesh generation techniques and methods, Ansys ICEM CFD was used in this study. Ansys ICEM CFD offers in-depth mesh analysis and optimization tools, including smoothing, refinement, coarsening, and element type conversions. Fuel cell geometry was firstdecomposed into individual volume elements to run a numerical approximation, using governing equations and investigating only the adequate elements that represent the cell layers to minimize the computation time. This allowed the best trade-off between precision and computational time to be optimized. All meshes were run to 10,000 iterations to ensure convergence. The material properties and all model parameters were set to be identical to those reported by Arvay et al. [3]. The resolution ofthe Navier-Stokes transport equations is used to represent the three-dimensional fluid flow and heat transmission. The Navier-Stokes equations, often known as conservation or transport equations, are written as [16]:

$$\frac{\partial}{\partial t}\int_V \rho\phi \, dV + \oint_A \rho\phi \mathbf{V}\cdot d\mathbf{A} = \oint_A \Gamma_\phi \nabla\phi\cdot d\mathbf{A} + \int_V S_\phi \, dV \qquad (4)$$

In the boundary condition in the cathode, the species mass fraction was set at 0.2 for oxygen and at0.1 for water. The species mass fraction of oxygen in the cathode was set at 0.9 for oxygen and at
0.1 for air. The inlet mass flow rate of the anode was established with a mass fraction of 0.2 for water and 0.8 for hydrogen. The temperature of the cell was set at 60°C. The operating condition was set at 301,325 Pa pressure.

Table 1. Model parameters.

Parameter	Units	Value
BP density	kg/m^3	1990
BP specific heat capacity	J/(kg K)	710
BP thermal conductivity	W/(m K)	120
BP electric conductivity	1/(Ω m)	92,600
BP thickness	mm	9.5
BP GDL contact resistance (parallel)	Ω m^2	4.56E-06
GDL density	kg/m^3	321.5
GDL porosity	e	0.82
GDL electric conductivity	1/(Ω m)	280
GDL viscous resistance (anode)	1/m^2	1.00E+12
GDL viscous resistance (cathode)	1/m^2	3.86E+12
GDL wall contact angle	deg	110
GDL thickness	μm	420
CL surface-to-volume ratio	m^2/m^3	1.25E+07
CL thickness (anode)	μm	6
CL thickness (cathode)	μm	12
Membrane density	kg/m^3	1980
Membrane thermal conductivity	W/(m K)	0.16
Membrane equivalent weight	kg/kmol	1100
Membrane thickness	μm	175
Open circuit voltage	V	0.98
H_2 diffusivity	m^2/s	8.00E-05
O_2 diffusivity	m^2/s	2.00E-05
H_2O diffusivity	m^2/s	5.00E-05
Pore blockage saturation exponent	e	2
Concentration exponent (anode)	e	0.5
Concentration exponent (cathode)	e	1
Charge transfer coefficient (anode)	e	1
Charge transfer coefficient (cathode)	e	1
Reference exchange current density (anode)	A/m^2	4.48E+05
Reference exchange current density (cathode)	A/m^2	4.48
Electrolyte Projected Area	m^2	0.0049

4. Results and discussion

In the cross-section mesh sensitivity analysis, three meshes with significantly different refinementlevels were used to exhibit mesh sensitivity of the entire cell. The first mesh was relatively small, with16,400 elements. The second mesh had 70,400 parts, and the third one incorporated 262,400 elements. The smallest mesh was cut with 20 cells horizontally, facing the cell, as Figure 3 shows. The mesh sensitivity analysis results of the above three sets of meshes and their normalization difference in current density are shown in Figure 4 (a). The results of the study show noteworthy differences between the small mesh and finer meshes. Comparing Figures 4 (b) through (d) we see that the small mesh (20 cells) is insufficient to establish a reasonable convergence rate. The result was more than 5% from the final value, but the rate of convergence was slowing down, as Figure 4 (b) illustrates.

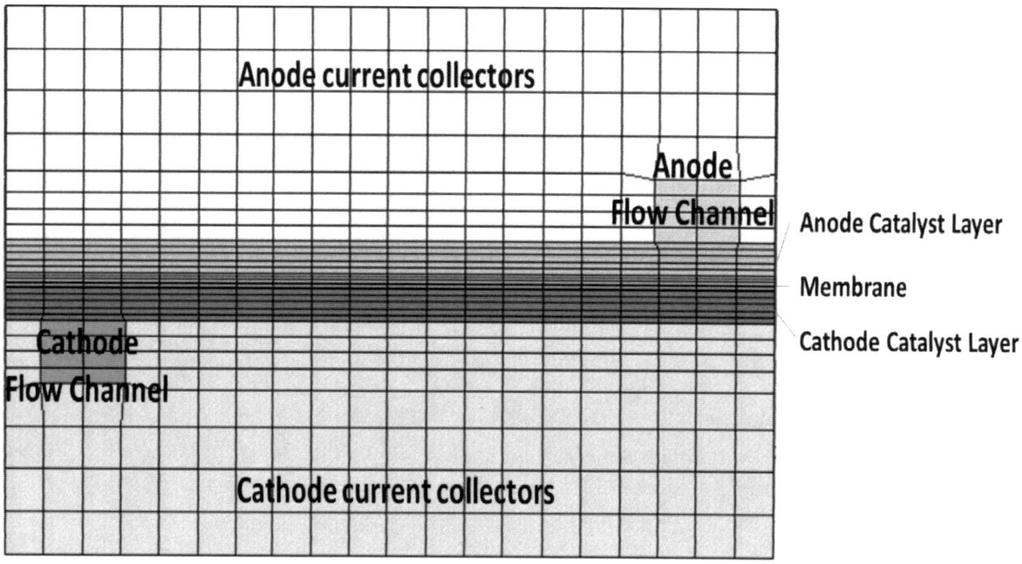

Figure 3. Mesh cross-section: the entire cell was divided into 20 cells horizontally.

Cell performance for 40 and 80 cells mesh at various voltages are shown in Figures 4 (c) and (d).The 40 cell mesh and 80 cell mesh results provided the same numerical result, but both of these meshesreached their final value after 8,000 iterations. This convergence history is satisfactory, and further iteration is not necessary. Since the convergence outcomes between these two meshes are constant, it isrecommended to use the 40 cell mesh to perform further mesh sensitivity analyses. The convergence time for this mesh was reasonable with average computing power, but a larger mesh would increase this computing time.

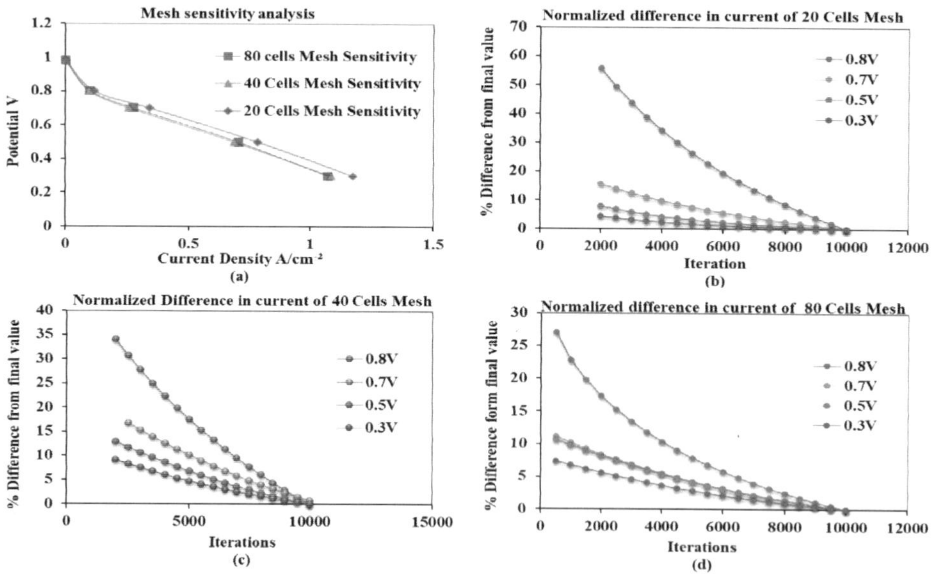

Figure 4. Mesh sensitivity convergence.

The subsequent analysis focused on the membrane electrolyte assembly (MEA) components and catalyst layers. This mesh aims to perform a membrane, catalyst layer (anode), and catalyst layer (cathode) mesh sensitivity analysis. The mesh sensitivity of the membrane is shown first to help improve the fundamental understanding of the mechanism of the membrane, including structure and thickness and the reactions that affect the fuel cell's performance. The diameter of the membrane should be kept as small as possible while being durable and providing excellent internal support. The thickness of the membrane layer is usually 50 to 175 microns. The width and temperature both affect membrane durability and performance. Therefore, the thickness should be in the tolerance and acceptable range for effective membrane performance and operation. The temperature should not exceed 100°C to ensure stability and permit hydration and performance optimization. In this mesh, three different numbers of vertical cells in the membrane were created. In contrast, the numbers of cells in both catalyst layers were constant to analyze the sensitivity. Table 2 below shows the varying number of vertical cells in the membrane, while holding the catalyst layer at a constant number. Secondly, the catalyst layer (anode) has meshed. As can be seen from Table 3, three different numbers of vertical cells in the catalyst (anode) were initiated to analyze their mesh sensitivity. Lastly, the mesh sensitivity analysis of the catalyst layer (cathode) was conducted. In this layer, 5, 10, and 20 vertical cells were generated while the membrane and anode held constant cell divisions.

Table 2. Mesh layers variation.

Anode	Membrane	Cathode
5	5	5
5	10	5
5	20	5

Each of these meshes was generated using the same set of parameters and properties attributes as is specified in Table 1. The operating temperature of these meshes was set at 60°C. The operating pressures were set at 201,325 Pa. The anode mass fraction was set at 0.5/0.5 H_2 to H_2O massratio. The cathode mass fraction was a 0.85/0.15 ratio of O_2 and H_2O, representing 100% of relative humidity. Four curves were generated at 0.8V, 0.7V, 0.5V, and 0.3V to determine the cell behavior and the effect of current density on convergence. The result of varying the number of cells in the membrane while the catalyst layer held constant is illustrated in Figure 5 (a). These results indicate that within the investigated range, varying the number of cells within the membrane makes no significant difference in the prediction of cell performance. The sensitivity analysis of the geometry of this cell shows that keeping the catalyst layer constant whilevarying the number of vertical cells in the membrane made no difference. This means that each membrane division can represent the membrane layer since these divisions produced the same result and converged in the solution simultaneously. On the other hand, as the number of divisions in the membrane increases, the overall current density increases as well, and as the number of divisions decreases, the overall current density reduces as well, which is consistent with what was expected. This indicates that the solution is converged. The same result was discovered in another computer simulation conducted by [3] that varying the membrane divisions makes no difference when all of the mesh parameters in the current collector and flow channels are kept constant.

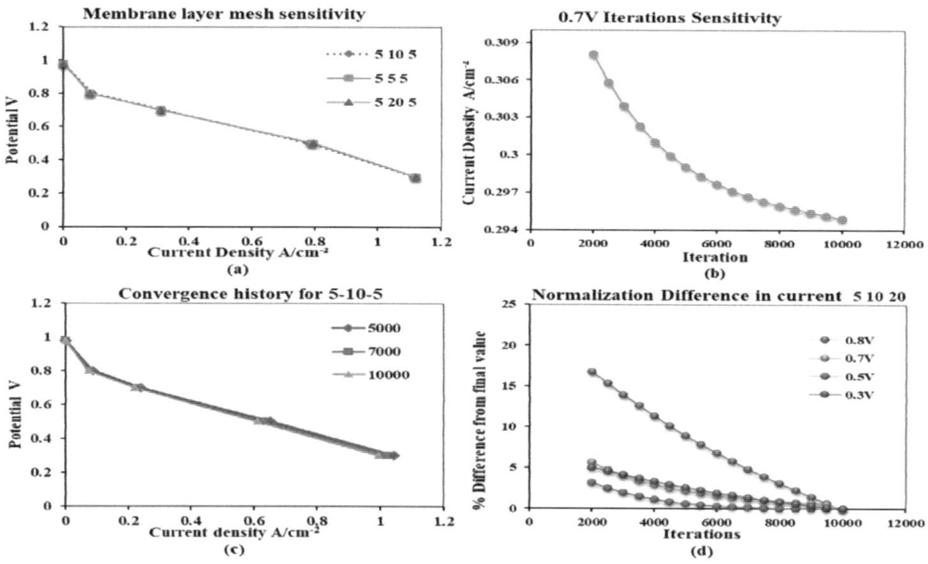

Figure 5. Membrane sensitivity analysis.

Furthermore, to determine the effect of current density on convergence, an iteration sensitivity test of 0.7V was generated, as Figure 5 (b) shows. This graph indicates that the data points merged and joined as the number of iterations increased. The normalized difference from the final value for different cell voltage values and membrane convergence history is also shown in Figures (c and d). This data suggested the number of iterations reached its final value and confirmed that 10,000 iterations per data point are sufficient, and there is no need for further iterations.

The catalyst layer (anode and cathode) is one of the essential components in the proton exchange membrane and is the core of the fuel cell energy conversion mechanism. This layer must be electronically and ionically conductive, gas permeable, and durable. Typically, the catalyst is platinum-based, with thickness in the range of 5 to 30 microns [29]. This layer transports electrons and protons through the circuit and it transports fuel and reactants to the reaction sites, promotes the reaction kinetics, and transports water to the gas diffusion electrodes or the membrane.

Firstly, a normalized difference from the final value for different cell voltage values (0.7V, 0.5V, and 0.3V) was generated to determine the effect of the number of iterations and cell convergence behaviors, as Figure 6 (a) displays. The result of this model shows that the solution in all data points of the catalyst (anode) reached the converged point at approximately 9,000 iterations. The data also shows that the lower voltage (0.3V) converges before the higher voltage (0.7V) simulations converge. Secondly, thecatalyst layer (anode) sensitivity simulation of varying the number of cells in this layer while holdingthe membrane layer and cathode layer constant was performed, as Table 3 shows. When the membrane and the cathode layer are held constant and vary the number of an anode layers, the results indicate that varying the number of computational cells within the anode makes no difference. Its values are consistent, as depicted in Figure 6 (b).

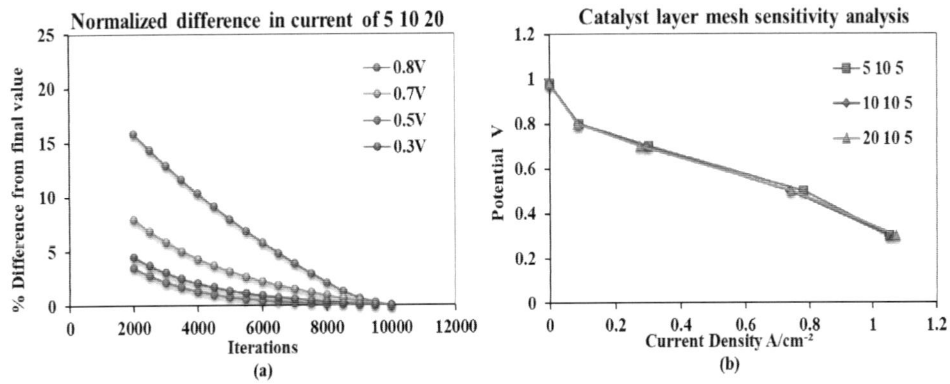

Figure 6. Anode mesh sensitivity analysis.

Table 3. Varying the cell divisions of catalyst (anode).

Anode	Membrane	Catalyst (cathode)
5	10	5
10	10	5
20	10	5

However, varying the cathode layer divisions while holding the membrane and anode layer consistent, as Table 4 shows, does make a significant difference. The solution of the catalyst (cathode) revealed that it did not converge completely and might need more iterations. In particular, the high voltage data points (0.8V) required more than 10,000 iterations to ensure convergence, as illustrated in Figure 7 (a). The iteration sensitivity study at higher current densities (> 0.8 A/cm^{-2}) also shows variance with lower current densities in the rate of convergence, as Figure 7 (b) indicated. This data suggests that the variation of the cathode layer could influence the fuel cell performance significantly. However, the mesh sensitivity analysis of this layer should be verified by additional experiments that show the humidity, water balance, and electrochemical reactions and thus provide a comparison with these predictions.

Table 4. Varying the cell divisions of catalyst (cathode).

Catalyst (anode)	Membrane	Catalyst (cathode)
5	10	5
5	10	10
5	10	20

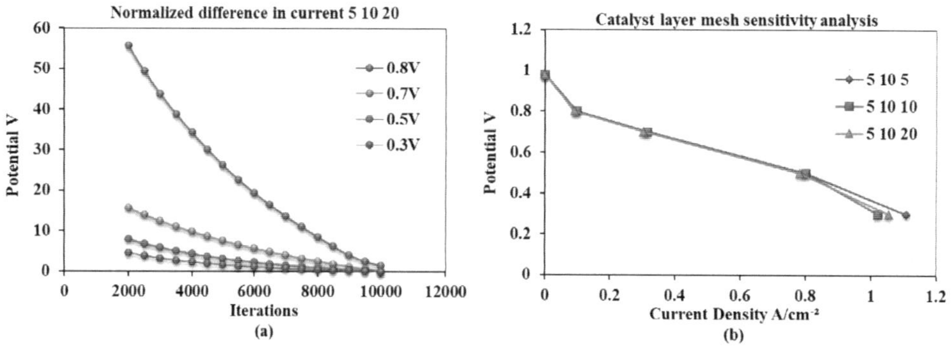

Figure 7. Cathode mesh sensitivity analysis.

5. Conclusion

Mesh sensitivity analysis in the ANSYS simulation study of a PEMFC was performed by varying the mesh density of membrane, anode, and cathode mesh. The outcomes of the present study showed that 40 cells are the minimum number of cells required to get a convergent solution in the cross-face direction of the PEM fuel cell. Also, it is anticipated that around 9,000 iterations will be required to achieve a precision of 5 percent of the value, which is a tolerable normalized difference between the first data point of 2000 and last point of approximately 12000 iterations. Results also indicated that 10,000 iterations per data point are necessary to ensure convergence to the acceptable error of 5%, which will increase the computation time. Varying the number of computational cells within the catalyst layer (anode) and membrane showed no significantdifference in the convergence. However, varying the catalyst layer (cathode) did make a significant difference. The five divisions in the membrane and catalyst layers (anode) are sufficient and represent cell layers. The membrane and anode data points of both layers converged at 9,000 iterations. Varying, decreasing, or increasing the catalyst layer (cathode) plays a significant role in determining PEMFC performance and mesh sensitivity. Its effect on the cell's overall performance was substantial and required a careful selection of cell divisions based on operating properties.

References

[1] Sun P. Modeling studies and efficient numerical methods for proton exchange membrane fuel cell. Computer Methods in Applied Mechanics and Engineering. 2011. pp. 3324–3340. doi:10.1016/j.cma.2011.08.007

[2] Li H, Knights S, Shi Z, Van Zee JW, Zhang J. (2010), Proton Exchange Membrane Fuel Cells: Contamination and Mitigation Strategies. CRC Press;

[3] Thangavelautham J. (2018). Degradation in PEM Fuel Cells and Mitigation Strategies Using System Design and Control. Proton Exchange Membrane Fuel Cell. doi:10.5772/intechopen.72208

[4] Arvay A, Ahmed A, Peng X-H, Kannan AM. (2012) Convergence criteria establishment for 3D

simulation of proton exchange membrane fuel cell. International Journal of Hydrogen Energy. Vol.37(3), pp. 2482–2489. doi:10.1016/j.ijhydene.2011.11.005

[5] Iranzo A, Muñoz M, Rosa F, Pino J. (2010) Numerical model for the performance prediction of a PEM fuel cell. Model results and experimental validation. International Journal of Hydrogen Energy. Vol 35(20), pp. 11533–11550. doi:10.1016/j.ijhydene.2010.04.129

[6] Cheddie DF, Munroe NDH. (2006) Three-dimensional modeling of high temperature PEM fuel cells. Journal of Power Sources. vol. 167(1), pp. 215–223. doi:10.1016/j.jpowsour.2006.01.035

[7] Siegel C. (2008). Review of computational heat and mass transfer modeling in polymer-electrolyte-membrane (PEM) fuel cells. Energy. vol. 33(9), pp. 1331–1352. doi:10.1016/j.energy.2008.04.015

[8] Kuo J-K, Yen T-H, Chen C-K. (2008). Three-dimensional numerical analysis of PEM fuel cells with straight and wave-like gas flow fields channels. Journal of Power Sources. vol. 177(1), pp. 96–103. doi:10.1016/j.jpowsour.2007.11.065

[9] Bednarek T, Tsotridis G, (2017) Issues associated with modelling of proton exchange membrane fuel cell by computational fluid dynamics, Journal of Power Sources, vol. 343, pp. 550-563. doi.org/10.1016/j.jpowsour.2017.01.059.

[10] D'Adamo A., Haslinger M., Corda G., Höflinger J., Fontanesi S., Lauer T. (2021) Modelling methods and validation techniques for CFD simulations of PEM fuel cells, Processes, vol. 9 (4), art. no. 688. doi: 10.3390/pr9040688

[11] Pant L.M., Mitra S.K., Secanell M., (2013) A generalized mathematical model to study gas transport in PEMFC porous media, International Journal of Heat and Mass Transfer, vol. 58(1-2), Issues, pp. 70-79. doi.org/10.1016/j.ijheatmasstransfer.2012.11.023.

[12] Gößling S., Nickig N., Bahr M., (2021) 2-D + 1-D PEM fuel cell model for fuel cell system simulations, International Journal of Hydrogen Energy, vol. 46(70), pp. 34874-34882. doi.org/10.1016/j.ijhydene.2021.08.044.

[13] Kone J.P, Zhang X., Yan Y., Hu G., Ahmadi G., (2018) CFD modeling and simulation of PEM fuel cell using OpenFOAM, Energy Procedia, vol. 145, pp. 64-69. doi.org/10.1016/j.egypro.2018.04.011.

[14] Wilberforce T, Khatib FN, Ijaodola OS, Ogungbemi E, El-Hassan Z, Durrant A, et al. Numerical modelling and CFD simulation of a polymer electrolyte membrane (PEM) fuel cell flow channel using an open pore cellular foam material. Sci Total Environ. 2019;678: 728–740.

[15] Ren, Xiaoming, Thomas E. Springer, Thomas A. Zawodzinski, and Shimshon Gottesfeld. 2000. "Methanol Transport Through Nation Membranes. Electro-Osmotic Drag Effects on Potential Step Measurements." Journal of The Electrochemical Society. https://doi.org/10.1149/1.1393219.

[16] Natarajan, Dilip, and Trung Van Nguyen. 2001. "A Two-Dimensional, Two-Phase, Multicomponent, Transient Model for the Cathode of a Proton Exchange Membrane Fuel Cell Using Conventional Gas Distributors." Journal of The Electrochemical Society. https://doi.org/10.1149/1.1415032.

[17] Wang, Yun, Suman Basu, and Chao-Yang Wang. 2008. "Modeling Two-Phase Flow in PEM Fuel Cell Channels." Journal of Power Sources. https://doi.org/10.1016/j.jpowsour.2008.01.047.

[18] Shah, A. A., K. H. Luo, T. R. Ralph, and F. C. Walsh. 2011. "Recent Trends and Developments in Polymer Electrolyte Membrane Fuel Cell Modelling." Electrochemical Acta. https://doi.org/10.1016/j.electacta.2010.10.046.

[19] Xia, Zhifeng, Bowen Wang, Zirong Yang, Kangcheng Wu, Qing Du, and Kui Jiao. 2019. "Effect of Operating Conditions on Performance of Proton Exchange Membrane Fuel Cell with Anode Recirculation." Energy Procedia. https://doi.org/10.1016/j.egypro.2019.01.428.

[20] Meng, Xiangchao, Hong Ren, Xiaokang Yang, Tienan Tao, and Zhigang Shao. 2022. "Experimental Study of Key Operating Parameters Effects on the Characteristics of Proton Exchange Membrane Fuel Cell with Anode Recirculation." Energy Conversion and

Management. https://doi.org/10.1016/j.enconman.2022.115394.

[21] Wang Y, Seo B, Wang B, Zamel N, Jiao K, Adroher XC. Fundamentals, materials, and machine learning of polymer electrolyte membrane fuel cell technology. Energy and AI. 2020. p. 100014. doi:10.1016/j.egyai.2020.100014

[22] Peighambardoust SJ, Rowshanzamir S, Amjadi M. Review of the proton exchange membranes for fuel cell applications. International Journal of Hydrogen Energy. 2010. pp. 9349–9384. doi:10.1016/j.ijhydene.2010.05.017

[23] Wang Y, Diaz DFR, Chen KS, Wang Z, Adroher XC. Materials, technological status, and fundamentals of PEM fuel cells – A review. Materials Today. 2020. pp. 178–203. doi:10.1016/j.mattod.2019.06.005

[24] Jaralla R. Proton exchange membrane (PEM) fuel cell parametric study via mathematical modeling and numerical simulation. doi:10.32920/ryerson.14649780

[25] Administrator. Types of Fuel Cells. [cited 28 Mar 2022]. Available: https://www.ceb.cam.ac.uk/research/groups/rg-eme/Edu/fuelcells/types-of-fuel-cells

[26] Tian P, Liu X, Luo K, Li H, Wang Y. Deep learning from three-dimensional multiphysics simulation in operational optimization and control of polymer electrolyte membrane fuel cell for maximum power. Applied Energy. 2021. p. 116632. doi:10.1016/j.apenergy.2021.116632

[27] Luo Y, Jiao K. Cold start of proton exchange membrane fuel cell. Progress in Energy and Combustion Science. 2018. pp. 29–61. doi:10.1016/j.pecs.2017.10.003

[28] Zhang G, Jiao K. Multi-phase models for water and thermal management of proton exchange membrane fuel cell: A review. Journal of Power Sources. 2018. pp. 120–133. doi:10.1016/j.jpowsour.2018.04.071

IOP Conf. Series: Materials Science and Engineering 1278 (2023) 012015 doi:10.1088/1757-899X/1278/1/012015

Optimum Configuration of Rainwater Harvesting System using CREO PARAMETRIC and ANSYS

MD Helal Uddin[1], Nurul Asma Samsudin[2], Nowshad Amin[2]
[1] Department of Mechanical Engineering, Universiti Tenaga Nasional (The Energy University), Jalan IKRAM-UNITEN, 43000 Kajang, Selangor, MALAYSIA.
[2] Institute of Sustainable Energy, Universiti Tenaga Nasional (The Energy University), Jalan IKRAM-UNITEN, 43000 Kajang, Selangor, MALAYSIA.
Corresponding Author: mdhelaluddinarnob@gmail.com, nowshad@uniten.edu.my

ABSTRACT

The practice of rainwater harvesting (RWH) has been studied extensively in recent years, as it has the potential to alleviate the increasing stress on urban water distribution systems. An appropriate design and evaluation of an RWH system is necessary to improve system performance and the stability of the water supply. In this study, the design of the main components in the RWH system such as the gutter, water storage tank and water pump were varied based on the main material of the component and rainfall intensity. Each component was presented according to the results of design, stress and fluid analyses performed by Creo parametric 7.0 and ANSYS. The main outcome of this work is a computational design tool that can be used to improve the RWH system component's performance considering a variety of factors. A design of a gutter made of polyvinyl chloride (PVC) with a yield strength of 50 MPa and a size of 5 inches was used. The rainwater will pass through a water pipe to a 500 L, High-Density Polyethylene (HDPE) water storage tank with a yield strength of 43 MPa. A 1 HP single-phase pump with a maximum flow rate of 196.2 L/min was used to generate a steady flow for household usage. The factor of safety of each component was obtained above 1.5 and was determined by the maximum Von Mises stress. Each component in the system has a different life span, however, the overall life span of the RWH system is estimated to be 20 years

Keywords: *Rainwater harvesting system, CREO parametric, ANSYS,* Von Mises stress

1. Introduction

Various parts of the world are affected by climatic variability which is associated with problems of water scarcity affecting entire populations. From a biological perspective of the living organism, water is a necessity for human life. It also serves as the foundation for food security and biodiversity [1]. Rain is the ideal source of recharge for the groundwater. However, the groundwater has limitations. Currently, groundwater levels are constantly and perpetually depleting as a result of over-extraction and injudicious usage [2]. Conventional water management and efficient rainwater storage can help alleviate water shortages and the pressure on water supplies by providing water for both domestic and commercial uses as well as agriculture. Rainwater harvesting is the deposition of rainwater for reuse on-site rather than runoff. By preventing runoff from entering the drainage system and reducing overloads on water treatment facilities, it effectively utilises rainwater. In parallel to that, it aids in replenishing water into aquifers and enhances groundwater quality through dilution. According to the World Health Organization (WHO, 2008), collected rainwater typically has fewer pathogens than unprotected surface water [3]. Water conservation has become the need of the day. Overcoming the escalating water shortage is one of the century's greatest challenges. If rainwater harvesting is used more frequently, water shortages could be alleviated. The method of capturing rainwater is simple, inexpensive, and has a number of advantages. This method of increasing the supply of domestic water has been used traditionally throughout time. Where there is a lack of water supply, it is particularly effective [4]. Due to the temporal and geographic variability of rainfall, rainwater harvesting has been done for more than

4,000 years. One of the purest water sources is rainwater that has been collected directly from the source. A typical rainwater harvesting system includes a catchment, storage tank, first flush driver, filter , gutters, and downpipes [5]. Increasing aridity and the ensuing water shortage are two of the biggest issues facing agriculture today. Harvesting rainwater may offer a solution to this problem. One of the main issues that countries all over the globe have been working to address over the past few decades is the scarcity of freshwater. The current water crisis forces a new paradigm in the management of water resources, and the implementation of sustainable water supply solutions is crucial. Using cisterns, tanks, and underground check dams, rainwater harvesting (RWH) is the process of collecting incident raindrops on roofs and ground surfaces. RWH has been used for centuries by humans to meet their water supply needs, and it can still be a valuable practise in the future of urban water use because it lowers household water consumption costs and provides a variety of opportunities for agriculture. In fact, due to an improvement in water productivity, irrigation of rainfed crops through RWH could represent a good option to boost crop yields [6]. Owing to an increase in the cost of tap water over the past 20 years, RWH systems have been used much more frequently than ever. If rainwater is utilised for toilet flushing in apartment complexes, it can substitute 30–40% of the tap water. When rainwater is used for washing, cleaning, and irrigation of gardens, drinking water savings may be increased. The financial viability of using rainwater collection systems varies greatly between countries and regions [7]. More than 71% of the surface of the Earth is covered by water, making it the most prevalent and significant substance there. Only 2.5% of the total amount of water that exists on the earth's surface is fresh water. Human use, industry, and agriculture all make use of fresh water. Due to the rapid growth in population, urban, industrial, and agricultural development, there is a very high demand for water [8].

Ali and Jain, 2014, established a rainwater harvesting system for the College of Engineering, Teerthanker Mahaveer University, India. The system has three major components: the catchment area, the delivery system, and the storage reservoir. He reported that water-resistant roofing materials such as iron sheets, tiles, and cement were utilised. The harvested rainwater was used for cleaning the building floors, labs, and classes, as well as for horticultural purposes [4].

Unaini, 2017, reported that a typical rainwater harvesting system contains six vital components: roof catchment, gutters, and downspouts; leaf screens and roof washers; storage tanks; conveying; and water treatment. She mentioned a flow metre was installed at the downpipe's end to calculate the volumetric flow of water the rainwater harvesting system installed. The three primary usage of the collected rainwater are flushing toilets, watering lawns, and washing cars [9].

This research study aims at finding a solution for rainwater harvesting systems in rural areas for domestic usage. First of all, methodology for rainwater harvesting (RWH) have been established, considering the types of components used, such as roof catchment, gutter, water tanks and pump. Afterwards, the type of material to be assigned, average rainfall, and the most applicable design parameters for modelling of the system. The materials chosen for the water supply are considered excellent for this usage according to guidelines provided by SPAN (Suruhanjaya Perkhidmatan Air Negara) on uniform technical guidelines water reticulation and plumbing. The 3D modelling of each component was created at Creo Parametric, which has the most flexible modelling capabilities and the ability to provide static simulation analysis for this study. The flow analysis was generated at Ansys Fluent owing to its accuracy and solid results.

2. Methodology

A RWH system consist of comprises of components at various stages such as catchments, storage, and distribution. Figure 1 shows the schematic diagram of household RWH system which comprises of gutter, the connecting pipes, water tank and water pump. In this study, the components such as gutter, water tank and water pump were designed using Creo Parametric software. The stress, strain, fluid and failure analysis of these designed components were tested using Creo Parametric and Ansys fluent software.

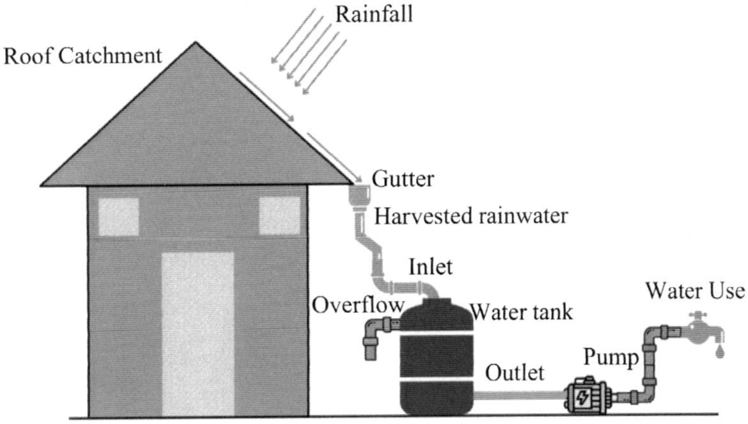

Figure 1: Schematic diagram of RWH system

2.1. Gutter

A CAD model of a trapezoid shape gutter was designed as shown in Figure 2 a-c. The dimensions of the model were listed in Table 1. In this model, 5 gutter brackets were used with distance of 0.17 m to hold the gutter to the catchment and provide ultimate grips and support during heavy rain.

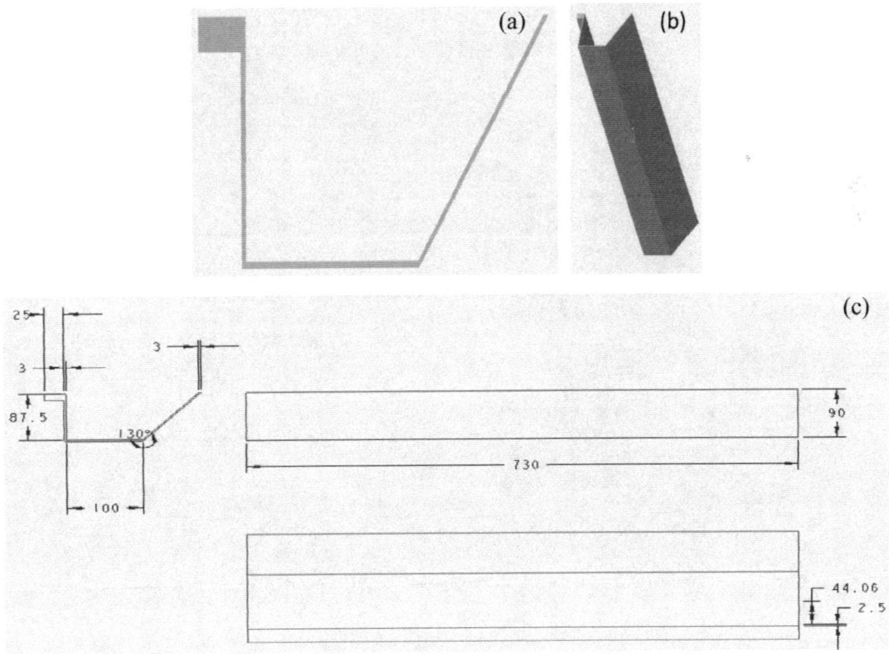

Figure 2: (a) Front view, (b) CAD model, and (c) dimensions of the gutter

IOP Conf. Series: Materials Science and Engineering 1278 (2023) 012015 doi:10.1088/1757-899X/1278/1/012015

Table 1: Dimensions of gutter.

Parameters	Values
Length	0.73 m
Shell thickness	0.0025 m
Height	0.0875 m
Wide	0.1 m
Base angle	130°

$$Volume\ of\ gutter = \frac{1}{2} \times (b1 + b2) \times h \times l = \frac{1}{2} \times (0.1 + 0.17292) \times 0.0875 \times 0.73 \quad (1)$$

Where, h is height of the gutter (0.0875 m), l is the length (0.73 m) b1 is base 1 (0.1m) and b2 is base 2 (0.17292 m).

$$Load\ exerted\ on\ gutter = \rho v g = (1000\ \times 0.0087163825\ \times 9.81) \quad (2)$$

Where ρ is the density of water (1000 kgm^{-3}) ,v is the volume of the gutter (0.0087163825 m^3), and g is the gravitational acceleration is (9.81 ms^{-2}), that's gives the load exerted on gutter is 85.50 N. Based on the CAD model, the estimated volume and the exerted load can be accepted by the gutter was calculated based on equation 1 and 2. The volume and the exerted load of the gutter obtained was 0.0087163825 m^3 85.5077 N, respectively. These values will be used for the stress, static and failure analysis of the gutter.

2.2. Water tank

A CAD model of water tank (Figure 3a and 3b) was designed based on the dimension listed in Table 2. One inlet, one outlet and one overflow pipe were used in this model.

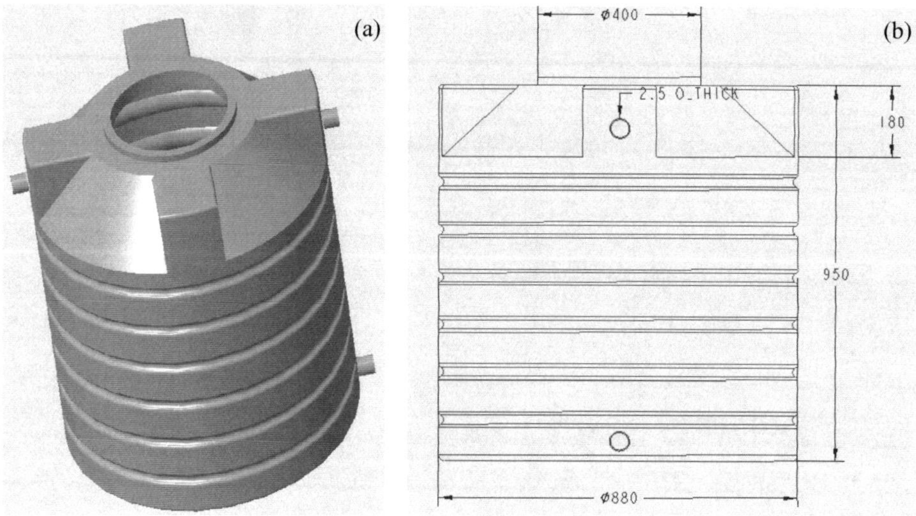

Figure 3: (a) CAD model, and (b) dimensions of water tank

Table 2: Dimensions of water tank.

Parameters	Values
Diameter	0.88 m
height	0.95 m
Manhole	0.4 m
Shell thickness	0.025 m
Inlet pipe, Outlet pipe, Overflow pipe	0.05 m

Based on the CAD model, the estimated volume, hydrostatic pressure, and overflow were calculated using the following equations:

$$Volume\ of\ water\ tank = \pi r^2 h = (3.1416 \times 0.4^2 \times 0.95) \quad (3)$$

Where, h is the height of the water tank (0.95 m), r is the radius which (0.44 m)

$$Hydrostatic\ pressure = h\rho g = (0.95 \times 1000 \times 9.81) \quad (4)$$

Where, ρ is the density of water (1000 kgm^{-3}), g is the gravitational acceleration is (9.81 ms^{-2}) and h is the height (0.95 m).

$$\text{Bernoulli's equation} = P_1 + \rho g h_1 + \frac{1}{2}\rho v_1^2 = P_2 + \rho g h_2 + \frac{1}{2}\rho v_2^2 = P_3 + \rho g h_3 + \frac{1}{2}\rho v_3^2 \quad (5)$$
$$= 101325 + (1000 \times 9.81 \times 0.95) + \frac{1}{2}(1000 \times 0) = 101325 + (1000 \times 9.81 \times 0) + \frac{1}{2}(1000 \times v_2^2) = 101325 + (1000 \times 9.81 \times 0.95) + \frac{1}{2}(1000 \times v_3^2)$$

Bernoulli's principle was applied for the equation 5 for measuring the overflow of the tank. P_1, P_2 and P_3 are the atmospheric pressure (101325 Pa), V_1 is velocity at the height of 0.95 m, V_2 is outlet velocity and V_3 is overflow velocity of the water. In accordance with Bernoulli's equation V_1 and V_2 lie on the same streamline with steady flow of water. The estimated volume of water tank (0.5778 m^3), the hydrostatic pressure (9319.5 Pa), the outlet velocity (4.316 ms^{-1}) and overflow velocity (-0.2507ms^{-1}) were then used in the further analysis.

2.3. Water pump

The centrifugal pump with a circular shape CAD model shown in Figure 4 (a-d) was designed using the dimensions shown in Table 3. The model was designed with an inlet pipe, an outlet pipe. and an impeller inside the pump. A spherical-shaped semi open impeller was designed with a total of six blades that were geometrically patterned for this modelling.

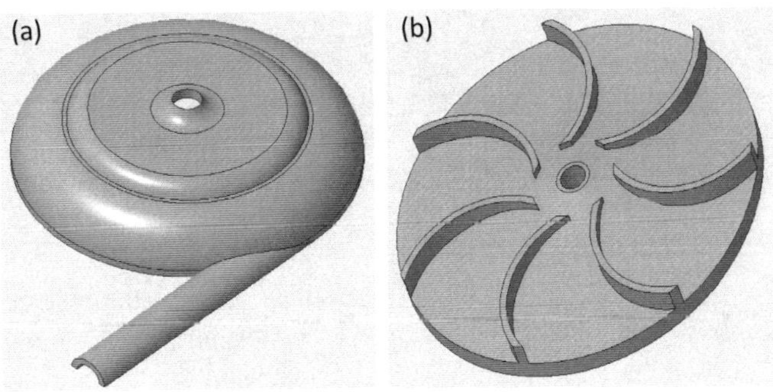

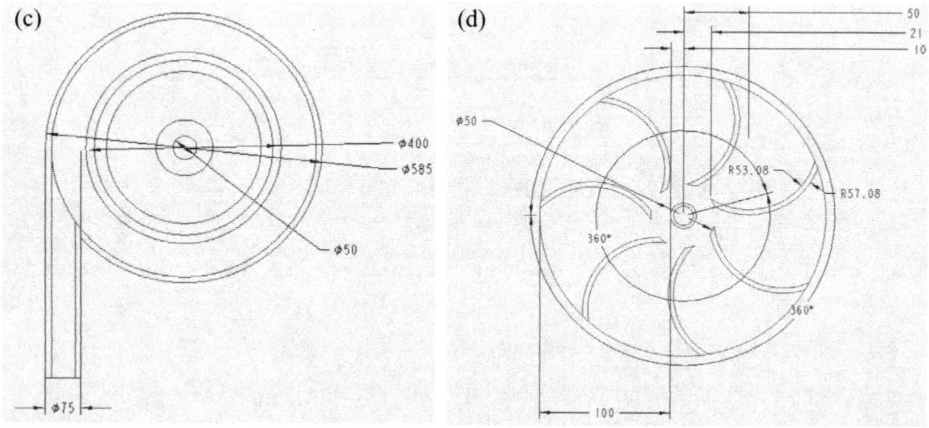

Figure 4: (a) CAD model pump, (b) CAD model impeller, (c) dimensions of pump, and (d) dimensions of impeller

Table 3: Dimensions of water pump.

Parameters	Values
Pump diameter	0.585 m
Inlet pipe	0.05 m
Outlet pipe	0.075 m
Outlet pipe length	0.5 m
Impeller diameter	0.22 m
Impellers revolve	0.1 mm
Impeller blade outer radius	0.05708 mm
Impeller blade inner radius	0.05308 mm

Figure 5 a-c shows the CAD model of assembled pump. The impeller was assembled and connected to the casing with a connecting rod.

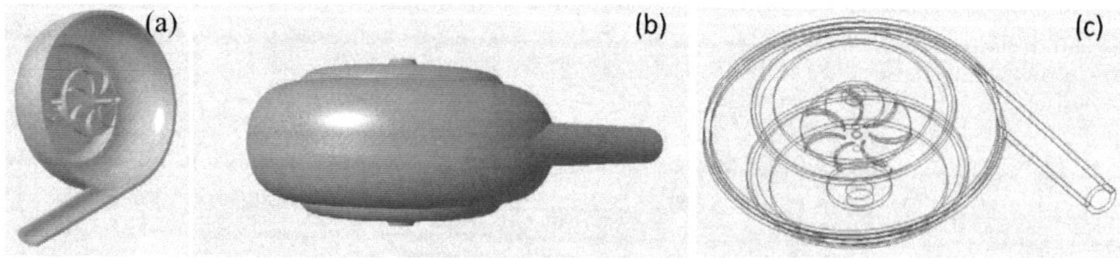

Figure 5: (a)Impeller assemble inside the pump, (b) assembled pump, and (c) assembled pump no shade view

The safety factor of the components was analysed based on the strength of the materials' yield and maximum von mises stress. In this study, the materials for each component were varied to observe the influence of the safety factor and obtained the best configuration of the RWH system. Materials such as polyvinyl chloride (PVC), unplasticized polyvinyl chloride (UPVC) and aluminium (Al) were tested for

IOP Conf. Series: Materials Science and Engineering 1278 (2023) 012015 doi:10.1088/1757-899X/1278/1/012015

optimum configuration for gutter and water tank while high-density polyethylene (HDPE), polyvinyl chloride (PVC) and stainless steel were tested for water tank.

The safety factor analysis for each component materials were calculated based on the following equation:

$$Factor\ of\ Safety = \frac{Yeild\ strength}{Maximum\ Von\ mises\ stress} \tag{6}$$

3. Results and Discussion

3.1. Gutter Static Analysis

The materials properties for PVC, UPVC and Al are presented in Table 4. The maximum von misses stress, strain at maximum principle, and strain energy per unit volume were analysed with applied load of 85.507 N.

Table 4: Gutter's materials properties.

Materials	Property	Values
Al	Young's Modulus	106×10^5 Psi
	Poisson's Ratio	0.33
PVC	Young's Modulus	435.1×10^3 Psi
	Poisson's Ratio	0.4
UPVC	Young's Modulus	414.8×10^3 Psi
	Poisson's Ratio	0.4

The static analysis of PVC gutter is shown in the Figure 6 (a-c).

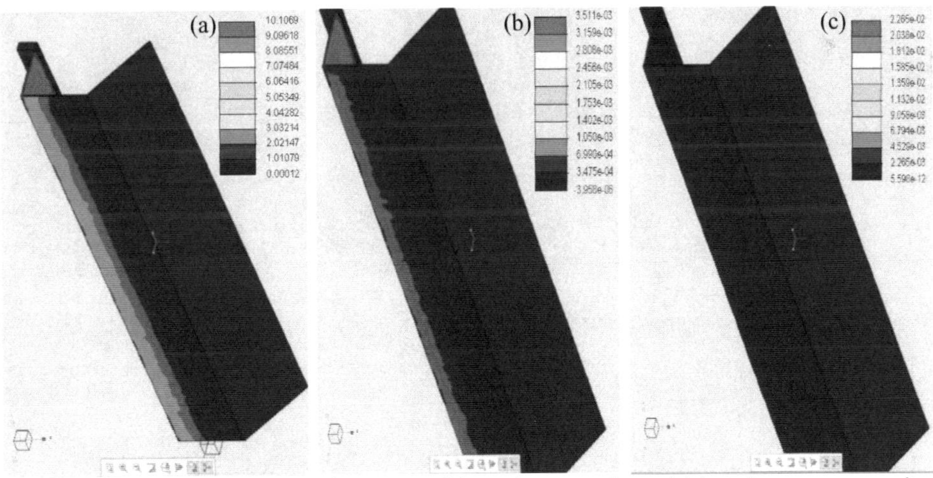

Figure 6: (a) Maximum von mises stress, (b) maximum strain, and (c) strain energy per unit volume

The static analysis of Al gutter is shown in the Figure 7 (a-c).

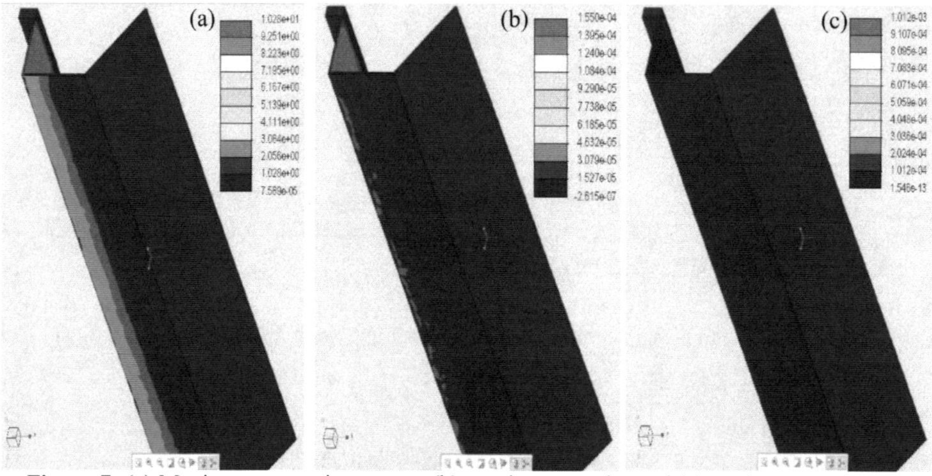

Figure 7: (a) Maximum von mises stress, (b) maximum strain, and (c) strain energy per unit volume

The static analysis of UPVC gutter is shown in the Figure 8 (a-c).

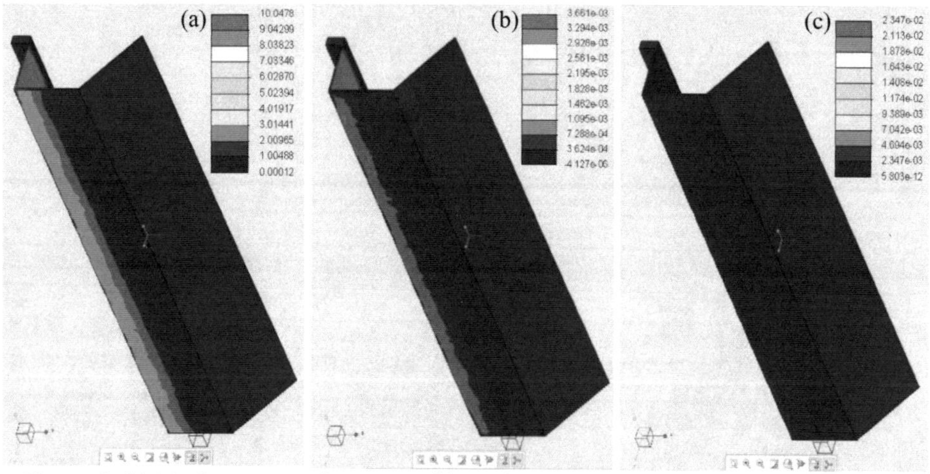

Figure 8: (a) Maximum von mises stress, (b) maximum strain, and (c) strain energy per unit volume

Table 5: Data analysis of gutter.

Materials	Applied Load (N)	Max Von Misses Stress (MPa)	Max Strain	Max Strain Energy Per Unit Volume	Factor of Safety	Yield Strength (MPa)
PVC	85.5	10.1077	3.51×10^{-3}	2.26×10^{-2}	5.310	53.7
AL	85.5	10.2800	1.55×10^{-4}	1.01×10^{-3}	40.27	414
UPVC	85.5	10.0478	3.66×10^{-3}	2.34×10^{-2}	4.970	50.0

A 5-inch gutter was used for the static analysis. The results of the investigation showed that all of the safety factor of all materials analysed were exceed the values of the acceptable range, which demonstrating an exceptional yield strength and durability of the materials y. Al offers an optimal safety factor of 40.27 and a maximum yield strength of 414 MPa, compared to PVC (53.7 MPa), and UPVC (50 MPa). The von mises stress analysis reveals the amount of distortion energy present when the components are under load. Based on results in Table 4, all materials able to withstand pressure almost similar pressure. This proved that all materials are suitable to be use as gutter.

The maximum principal strain determines when the gutter was put under the most stress until it is breaking. The results show that UPVC has the highest maximum strain, 3.66×10^{-3}, followed by PVC at 3.51×10^{-3} and Al at 1.55×10^{-4}. The energy that is stored as a result of deformation is referred to as strain energy. The highest strain energy obtained was 2.26×10^{-2} MPa for PVC, 2.34×10^{-2} MPa for UPVC, and 1.01×10^{-3} MPa for Al. The load was equally distributed across the entire gutter surface. The constraint was applied at the edge of the top left of the gutter which will be connected to the roof catchment with support of gutter brackets. The maximum stress was observed at joint point. Every configuration exceeded the minimum safety factor of 1.5, and the designs were adequate for the system.

Although the factor of safety of all materials are high, PVC is considered as the best suited as gutter for RWH system due to its low cost, flexibility, availability, resistance to rust, and ability to withstand sunlight [10].

3.2. Water Tank Static Analysis

HDPE, PVC, and stainless-steel materials were used for water tank static analysis. The properties of the material are presented in Table 6. The maximum von misses stress, strain at maximum principle, and strain energy per unit volume were analysed using applied pressure of 0.00931 MPa.

Table 6: Water tank's material properties.

Materials	Property	Values
HDPE	Young's Modulus	156.6×10^{3} Psi
	Poisson's Ratio	0.418
PVC	Young's Modulus	435.1×10^{3} Psi
	Poisson's Ratio	0.400
Stainless steel	Young's Modulus	280.0×10^{5} Psi
	Poisson's Ratio	0.300

The static analysis of HDPE water tank is shown in the Figure 9 (a-c).

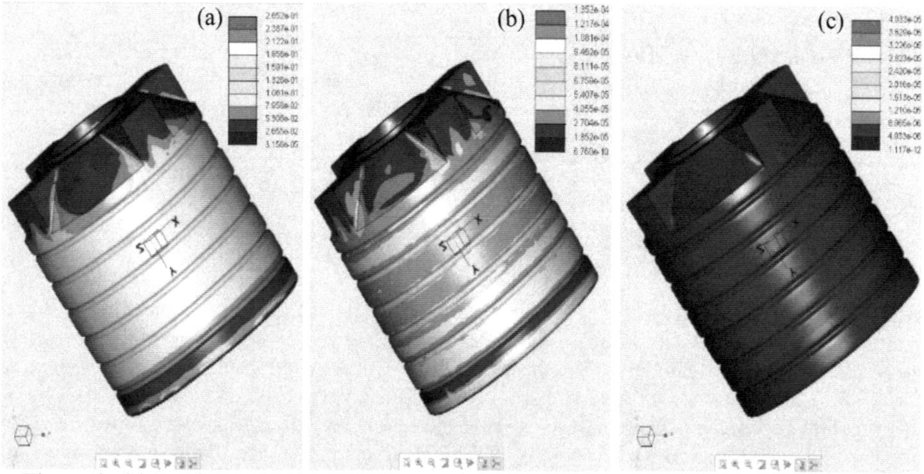

Figure 9: (a) Maximum von mises stress, (b) maximum strain, and (c) strain energy per unit volume

The static analysis of PVC water tank is illustrated in the Figure 10 (a-c).

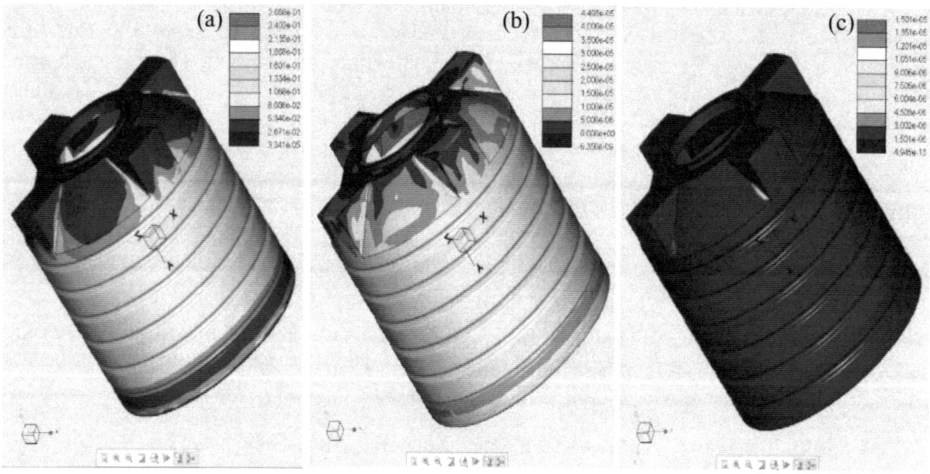

Figure 10: (a) Maximum von mises stress, (b) maximum strain, and (c) strain energy per unit volume

The static analysis of Stainless-steel water tank is demonstrated in the Figure 11(a-c).

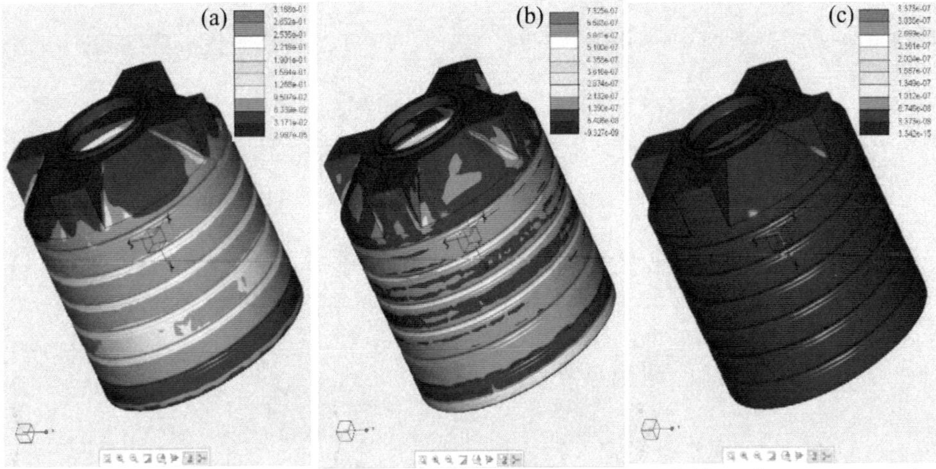

Figure 11: (a) Maximum von mises stress, (b) maximum strain, and (c) strain energy per unit volume

Table 7: Data analysis of water tank.

Materials	Pressure (MPa)	Max Von Misses Stress (MPa)	Max Strain	Max Strain Energy Per Unit Volume (MPa)	Factor of Safety	Yield Strength (MPa)
HDPE	0.00931	2.65×10^{-1}	1.35×10^{-4}	4.04×10^{-5}	1.62×10^{2}	43.0
PVC	0.00931	2.67×10^{-1}	4.49×10^{-5}	1.50×10^{-5}	1.99×10^{2}	53.7
Stainless Steel	0.00931	3.17×10^{-1}	7.74×10^{-7}	3.37×10^{-7}	6.40×10^{2}	205

The static analysis was done on a 500 L water tank. Table 7 shows the results for maximum von mises stress, maximum strain, maximum strain energy per unit volume, factor of safety and yield strength of the water tank. Based on the results, the safety factor of all materials is above the range, which indicates that these materials have excellent yield strength and durability. Stainless steel has a maximum yield strength of 205 MPa and provides an optimum safety factor of 6.40×10^{2} compared to HDPE (43 MPa) and PVC (53.7 MPa). The results also showed stainless steel have higher maximum von mises stress (3.17×10^{-1} MPa) compared to HDPE and PVC (2.65×10^{-1} MPa).

HDPE has a maximum strain value of 1.35×10^{-4}, followed by UPVC at 4.49×10^{-5}, and stainless steel at 7.74×10^{-7}. The strain energy refers to the energy stored due to the deformation. HDPE has a maximum strain energy of 4.04×10^{-5} MPa, followed by PVC at 1.50×10^{-5} MPa and stainless steel at 3.37×10^{-7} MPa. The constraint was applied at the bottom surface of the tank, and the pressure was equally distributed over the entire surface of the tank. All the configurations were above the factor of safety, and the designs were suitable for the system.

Based on the quality analysis, stainless steel gives a better performance followed by HDPE and PVC. However, both HDPE and PVC are inexpensive and suitable for the system. Nevertheless, considering on the yield strength of the material and its durability, HDPE was selected for this study.

3.3. Failure Analysis of the gutter and water tank

The failure analysis was conducted to determine the maximum load that can be apply on the gutter and maximum pressure on the water tank before its start to bend.

Table 8: Gutter failure analysis.

Material	Component	Max Applied Load	Max V.M Stress (MPa)	FOS
PVC	Gutter	298 N	35.226	1.5045

Table 9: Water tank failure analysis.

Material	Component	Max Applied Pressure	Max V.M Stress (MPa)	FOS
HDPE	Water Tank	1 MPa	28.4682	1.5357

Table 8 shows that a maximum 298 N load can be apply on the PVC gutter. The load above that will cause the gutter to bend and the system to fail. Moreover, that will be below factor of safety range. Similarly, a maximum pressure of 1 MPa is achievable for the water tank (Table 9).

3.4. Centrifugal water pump ANSYS fluent analysis

This stage of the modelling process is to create the meshing of the pump. The parameters are used for meshing and flow analysis are presented in Table 10. The model of the pump is shown in the Figure 12 (a-d).

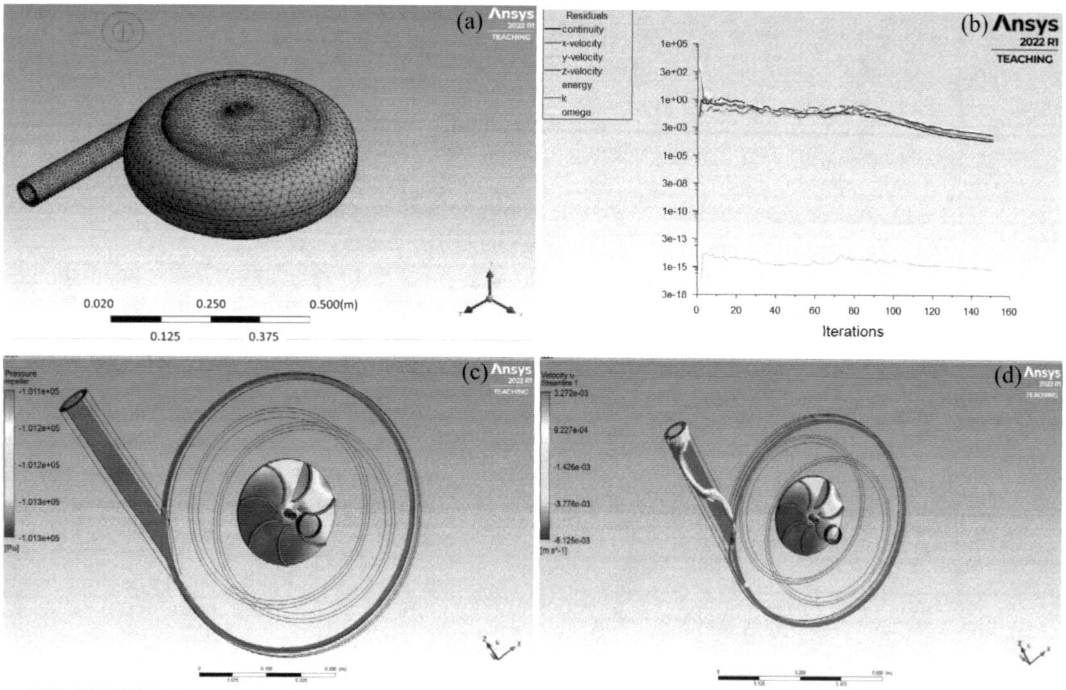

Figure 12: (a) Mesh matrix of pump, (b) Iteration, (c) pressure on the impeller, and (d) velocity streamline of the outlet

Table 10: Pump flow analysis.

Parameters	Values
Elements	95443
Nodes	24776
Impeller angular velocity	1450 rpm
Mass flow rate	78kg/s
Maximum velocity	3.27E-3 ms^{-1}
Impeller static pressure	-1.01×10^{-5} Pa
Flowrate	196.2 L/min

A total of 95443 elements with 24776 nodes were employed for the fluent analysis and 1000 iterations were studied to determine the optimum velocity and pressure of the pump. The impeller angular velocity was set to 1450 rpm to get the optimum outlet velocity and water pressure. Water was used as the as material to be investigated. The mass flow rate of 78 kg/s and hybrid initialization were used to study the configurations. The maximum velocity obtained was 3.27×10^{-3} ms^{-1}. That provides the flowrate of 196.2 L/min. The static pressure obtain for the impeller was -1.01×10^{-5} Pa. The required flowrate of the centrifugal pump is to provide steady flow is 110 L/min. The designed pump provides flow rate of 196.2 L/min that vastly superior.

4. Conclusion

Based on the modelling, simulation and the analysis carried out in this study, the fabrication of the gutter and water tank materials in the correct shape can help enhance durability and life. In this study, PVC gutter was considered the best material for gutter, considering the factors of safety of the material and economic value and its resistant to rust. For water tank, a good match was found in HDPE due to its excellent yield strength, low cost, and durability. Lastly, the centrifugal pump with 1450 rpm was analysed and found suitable for using RWH system for household usage. Future research could examine the different configurations of the composition and material quality. The system has the potential to be used for drinking purposes if a filter is installed.

References

[1] L. Martínez-Acosta, A. A. López-Lambraño, and A. López-Ramos, "Design criteria for planning the agricultural rainwater harvesting systems: A review," *Appl. Sci.*, vol. 9, no. 24, 2019, doi: 10.3390/app9245298.

[2] T. Vinaya, "Need of Rain Water Harvesting," *Agric. Environ. E-newsletter*, vol. 1, no. 2, pp. 95–98, 2020.

[3] I. Parajuli, "Potential and Practices of Rain Water Harvesting System in Potential and Practices of Rain Water Harvesting," in *2018 SMART WATER GRID INTERNATIONAL CONFERENCE 12- 13 November 2018, Incheon, Republic of Korea*, 2018, no. November.

[4] S. R. Ali and R. K. Jain, "Rain Water Harvesting System for College of Engineering, Teerthanker Mahaveer University, Moradabad," *Int. J. Innov. Res. Sci. Eng. Technol.*, vol. 03, no. 08, pp. 15649–15657, 2014, doi: 10.15680/ijirset.2014.0308077.

[5] P. Selvakumar, T. Harish, and S. Sakthi Murugan, "Study on the effects of rain water harvesting system in an urban area," *Water Energy Int.*, vol. 63r, no. 3, pp. 56–60, 2020.

[6] L. Pari, A. Suardi, W. Stefanoni, F. Latterini, and N. Palmieri, "Economic and environmental assessment of two different rain water harvesting systems for agriculture," *Sustain.*, vol. 13, no. 7, 2021, doi: 10.3390/su13073871.

[7] G. Sakson, "Cost analysis of a rainwater harvesting system in Poland," *E3S Web Conf.*, vol. 45, pp. 1–8, 2018, doi: 10.1051/e3sconf/20184500078.

[8] S. Panhalkar, "Domestic Rain Water Harvesting System : A Model for Rural Development," *Soc. Sci. Nat.*, vol. 2, no. 3, pp. 861–867, 2011.

[9] A. Mohd Unaini, "Assessment on the Performance of a Rainwater Harvesting System," *Sci. Res.*, vol. 5, no. 3, p. 36, 2017, doi: 10.11648/j.sr.20170503.13.

[10] R. B. İsmail, A. Reyhan, and S. Zeynep, "An investigation into the building-sanitary equipment in the traditional residential architecture," *Sci. Res. Essays*, vol. 9, no. 15, pp. 668–686, 2014, doi: 10.5897/sre2012.750.

Acknowledgement

Authors would like to express their gratitude to National Science Foundation for the support through grant code of 202001001NSF

Electronic energy benchmarking system (e-EBS) as sustainability tool for public buildings in Malaysia

Abdul Murad Zainal Abidin[1], Norhayati Mat Wajid[2*], Faiz Fadzil[2], Samsiah Omar[2], Syahrul Nizam Kamaruzzaman[3], Nik Elyna Myeda Nik Mat[3], Anissa Azlan[3]

[1]Centre of Excellence for Engineering and Technology (CREaTE), Public Works Department Malaysia, 78000 Alor Gajah, Melaka, MALAYSIA.

[2]Mechanical Engineering Branch, Public Works Department Malaysia, 50480 Kuala Lumpur, MALAYSIA

[3]Faculty of Built Environment, University Malaya, 50603, Kuala Lumpur, MALAYSIA

*Email : norhayatimat@jkr.gov.my

Abstract. The global increase in carbon emission from buildings is driving greater urgency towards carbon neutrality. A number of strategies have been developed worldwide that include energy management and building-integrated photovoltaics but these are implemented independently. There is a need for integrated approach towards better energy monitoring and benchmarking. The aim of the research is to develop an electronic energy benchmarking system (e-EBS) for public buildings in Malaysia as dashboard on energy consumption, carbon footprint, and green rating score with strategies for improvement. This is done by integrating energy auditing with the Penarafan Hijau JKR (pHJKR) building rating scheme, and the JKR Energy Online System (JENOSYS). The benchmarking system would analyse energy data and give the rating with recommended energy-saving strategies for building owners. The tool for analysis and benchmarking was selected by comparing the coefficient of determination (R^2) of regression model with multi-layer perceptron, a type of neural network. The neural network model was found to have a slightly higher sensitivity level (R^2=0.95) compared to the regression model (R^2=0.93) and selected to develop a standard baseline of energy consumption for buildings in e-EBS. The following step is the validation of e-EBS by creating more energy benchmarks based on building category.

Keywords: Benchmarking, Energy efficiency, Simulation, Dashboard, Neural network

IOP Conf. Series: Materials Science and Engineering 1278 (2023) 012016 doi:10.1088/1757-899X/1278/1/012016

1. Introduction

Energy monitoring has numerous societal benefits, regardless of whether the energy consumer is small or large. Energy monitoring or energy auditing allows consumers to identify the most significant potential energy savings, which can reduce energy costs and improve profitability [1]. An energy audit programme will help to focus on variations in energy costs and provide feasible solutions for energy management.

There are few online monitoring systems developed by the companies or government departments in Malaysia. However, the energy monitoring features in these systems are not directly linked to the current green rating tools in Malaysia. As a result, the energy monitoring system operates independently of the green building certification process. Consequently, this imposes a challenge to building owners who have to use several separate tools to do benchmarking. To tackle the challenge, the research attempts to bridge the gap between energy monitoring systems with the green building rating tools in Malaysia by developing an Electronic Energy Benchmarking System (e-EBS) for government buildings.

Considering computational advancement, producing energy analysis may have a wider outlook such as proposing a larger energy-saving potential. Identifying the intrinsic link between building energy consumption and the analysis of associated variables can be performed. Research at Tianjin was done on an occupancy-based model for building electricity consumption prediction [2].The research addressed the issue of the intrinsic link between the occupant and building energy consumption and provides evidence about the extent to which the case building electricity consumption is related to its occupants and can be benchmarked separately for the building area and number of occupants. The associated variables concluded in the prediction analysis were the framework of the electricity consumption model, occupancy rates in the building and the growth limit of the building electricity consumption.

In practice, the portion of the energy usage that is indirectly connected to occupant behaviour can be estimated using the area index process, whereas the remainder can be calculated using direct connections between real-time occupancy and hourly power rates of equipment in buildings. In the research at Tianjin, the case building electricity consumption was split into 'basic' and 'variable' parts where the basic electricity consumption is constant throughout the day while variable electricity consumption is what changes depending on the needs of the occupants. The splitting of data established the occupancy-based model, to quantify the consumption at both the occupant and building levels in the research.

A similar study to explore potential energy savings in hospitals were conducted in Malaysia in relation to achieving the highest green building rating possible where three methods were employed; namely energy audit, empirical evidence, and simulation works [3]. The IES<VE> simulation software were used where the energy saving potentials cannot be measured from energy audit.

In the fields of Advanced Neural Network (ANN), an investigation in the estimation of the electricity consumption through the ANN method was conducted at the Mara University of Technology (UiTM) for two laboratories that place heavy types of machinery and equipment [4]. The ANN method can model energy consumption for a random day and the next hours. ANN method also has the advantages such as the mapping function in the neural network is very flexible and neural network has the ability of generalizing form a limit set of data points and giving a good result at new data points. For prediction at the two heavy laboratories in UiTM, ANN using feedforward neural network type is applied. The input data are power consumption, number of experiments run, hours of experiment and electricity cost while the output data is energy consumption. For the development of the ANN, the input data are divided into 70/30 in terms of percentage of training and testing. The 70/30 ratio is to give the higher coefficient of correlation. Change in usage patterns and occupant awareness could significantly affect the amount of energy of a premise. With the usage of computational tool, huge opportunities could be led towards energy saving and improved energy efficiency. The prediction model done at UiTM is useful and can be further applied to other premise with the faculty for energy management.

IOP Conf. Series: Materials Science and Engineering 1278 (2023) 012016 doi:10.1088/1757-899X/1278/1/012016

An energy benchmarking analysis for Malaysia government hospitals was conducted at the University Malaysia Sabah (UMS) [5]. The methodology involved collection of appropriate data through hospital building energy assessment survey to 120 hospital buildings which gives emphasizing on the annual energy consumption (AEC) of hospital buildings in term of electrical energy use index (EEUI), distribution of major energy using systems and equipment and analysis of other significant energy consumption drivers. The data gathered from the hospital energy assessment survey was then used as input file creation of Energy Plus simulation, a building energy simulation modelling software. AEC prediction was done using Microsoft Excel Regression Analysis (RA) and the model was intended to represent all hospitals and emphasize on the variability of energy consumption. The developed model was evaluated using the square of the correlation coefficient, R^2 to relate how well the independent variables in predicting variations in the hospital building AEC and the residual of the regression model is used to test the overall significance of the equation.

In comparison of metered annual hospital energy use and the predicted value, the regression model proposed could clarify 85% of the annual energy consumption. From this study, it is recognized that attaining reliable data through the assessment survey was not sufficient. The authors highlighted the importance of data collection technique to provide better and comprehensive information for better modelling. A similar work on using RA for energy performance study on was done at Chiang Mai University, Thailand [6]. This study emphasized in energy utilization index (EUI) and the analysis of factors affecting energy use using Multiple Linear Regression Analysis. The preliminary data of 45 hospitals were collected as these hospitals were using more than 3,000kW of energy more than suggested in Energy Conservation and Promotion Act in Thailand and these hospitals have complete data of energy use from 2010 until 2015. The model proposed from the analysis can be used as the energy performance indicator for future energy benchmarking in hospital buildings that operate for 24 hours. These two studies are to show the adaptability of RA techniques in developing modules for energy benchmarking as well as the importance of data acquisition techniques in ensuring reliable modelling.

In the context of integration of RA and ANN, another element could also be included, which is data classification according to occupancy hours to predict hourly or sub-hourly energy usage in buildings, as demonstrated by Ridwana, Nassif, and Choi at different locations in the United States and with different operating hours [7]. It was also demonstrated that energy estimation using RA and ANN, compared to conventional regression model alone, gives higher accuracy.

Energy baseline is an established data used as a basis for assessment or control. It is applied for the selection of the most appropriate pathway and the selected pathway will be applied to create baseline sitting which is suitable for control model for building energy management. To make baseline sitting, it requires considering the variation in energy consumption and factors affecting energy use, capacity in monitoring and report of control model for building energy management. Additionally, energy benchmarking derived from data collection of buildings and its engineering services should satisfy a set of standards such as it must be reached by a large proportion of the buildings and apply to all similar buildings, the performance indicator on which the benchmark is based should be easy to quantify from statistical information and easy to understand. Benchmarks should be dynamic in that continuous analysis and improvement can be made to increase performance level as the building energy efficiency requirement improved.

2. Research framework

Regression analysis (RA) is the most used method for developing baseline energy models [8]. However, RA is less accurate in analysing the non-linear characteristic. As a result, ANN have been used to replace the traditional method in a number of works [9]. ANN is a popular forecasting technique that mimics the operation of the human brain. It has been used to solve a wide range of engineering problems [8]. Because ANN learns from examples, large amounts of training data are required to achieve the best results. Inadequate or small data sets typically produce inaccurate results and a large error during the training stage.

A web-based energy performance reporting tool called JENOSYS incorporates a database of utility account numbers acquired from the utility service provider for analysis and reporting. JENOSYS was developed by the Public Works Department Malaysia (PWD). The JENOSYS framework was designed to ease the user to view the result of energy performance of their building in terms of energy saving in kWh, energy saving in utility cost (RM) and the annual building energy intensity (BEI) in kWh/m2/year [10].

Improvising a well-established online monitoring and reporting tool is proposed by embedding Machine Learning platform following the green rating tools in Malaysia. Automating process involves employing Artificial Intelligence platforms that can support the digitalization process and deliver the same or better results those human beings would have achieved. Hence, the research output will be to bridge the gap between energy monitoring systems with the green building tools in Malaysia by developing an Electronic Energy Benchmarking System (e-EBS) for government buildings. The basic concept framework of e-EBS for public buildings is shown in figure 1.

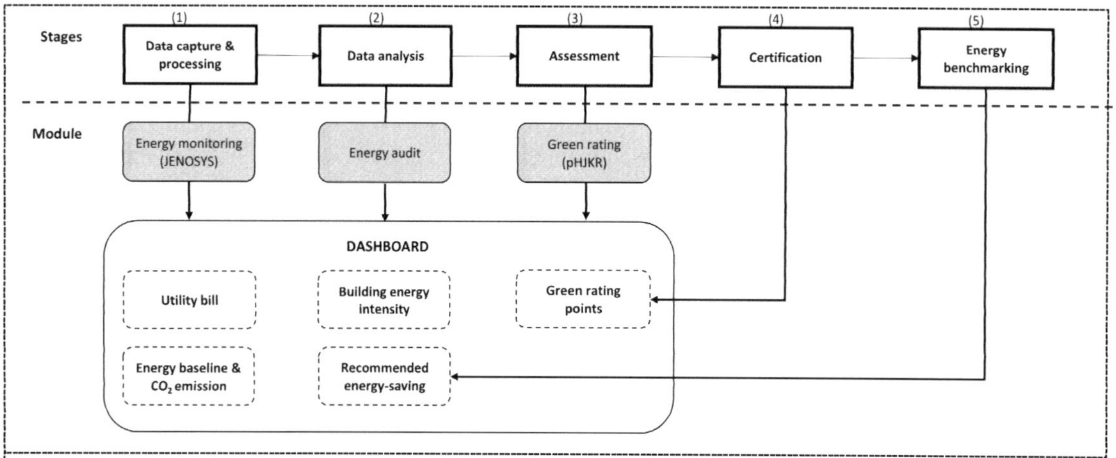

Figure 1: Framework for e-EBS.

The framework was divided into five (5) stages that start with data capture and processing (1), where the data on utility bills, energy used in kilowatt-hours (kWh) are extracted from the JENOSYS. The data analysis stage (2) calculates the data from the energy audit of a building. The assessment stage (3) is the usage of the green rating template that is related to energy use in buildings. The certification stage (4) is the evaluation of the rating points in order to award the star rating based on the pHJKR rating scheme. Lastly, the energy benchmarking stage (5) recommends the further energy-saving to achieve higher green rating, in addition to the savings that have been achieved.

The e-EBS is made up of three modules that automatically analyse a user building energy profile in accordance with pHJKR requirements. The integration of pHJKR into the system provides a more efficient method of assessment and documentation than the traditional and manual approach. This is stated to be possible because, after researching the pHJKR assessment guideline and submission requirements, a set of standard property sets will be identified as a pHJKR-rated template, which will then be included in the e-EBS database. Users would then be able to extract, interpret, and process data in accordance with the pHJKR guidelines. The system generates a report on conditional pHJKR points and recommended energy-saving strategies for further pHJKR certification evaluation. As a result, the system takes a comprehensive approach.

3. Methodology

The entire process flow of research is shown in figure 2, where the study starts with gathering the data from a building energy audit report, the pHJKR green building rating score, and the energy consumption from the JENOSYS monitoring tool. The results of workshops organised for building owners were acquired on the content of the dashboard for e-EBS. The next activity is the development of programming to integrate the data from the three components of e-EBS for the dashboard using an office building energy consumption data as a base case, and the final step is the validation of the e-EBS by using real data from buildings of different categories.

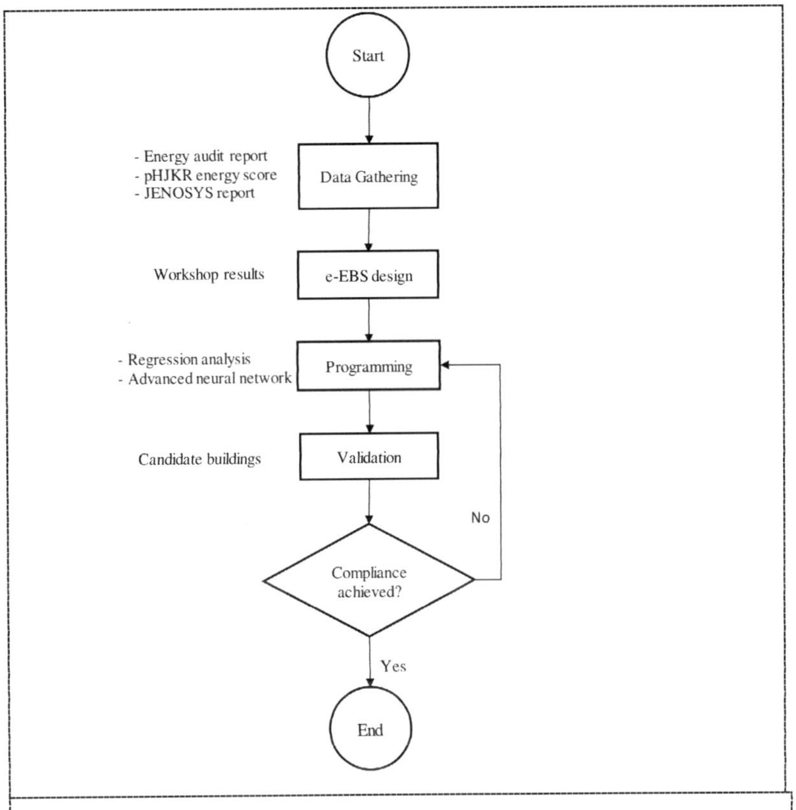

Figure 2. Flowchart of e-EBS development.

The determination of energy baseline is critical for building owners to set the target for reducing energy use in buildings, and the baseline would be subject to changes to the user behaviour, and the change in the building function, and also the change in the energy-related facility inside the building. The benchmarking process depends on the accuracy in the baseline determination. Therefore, the RA and ANN were compared in terms of sensitivity in order to calculate the baseline in the e-EBS, the process of which is shown in figure 3.

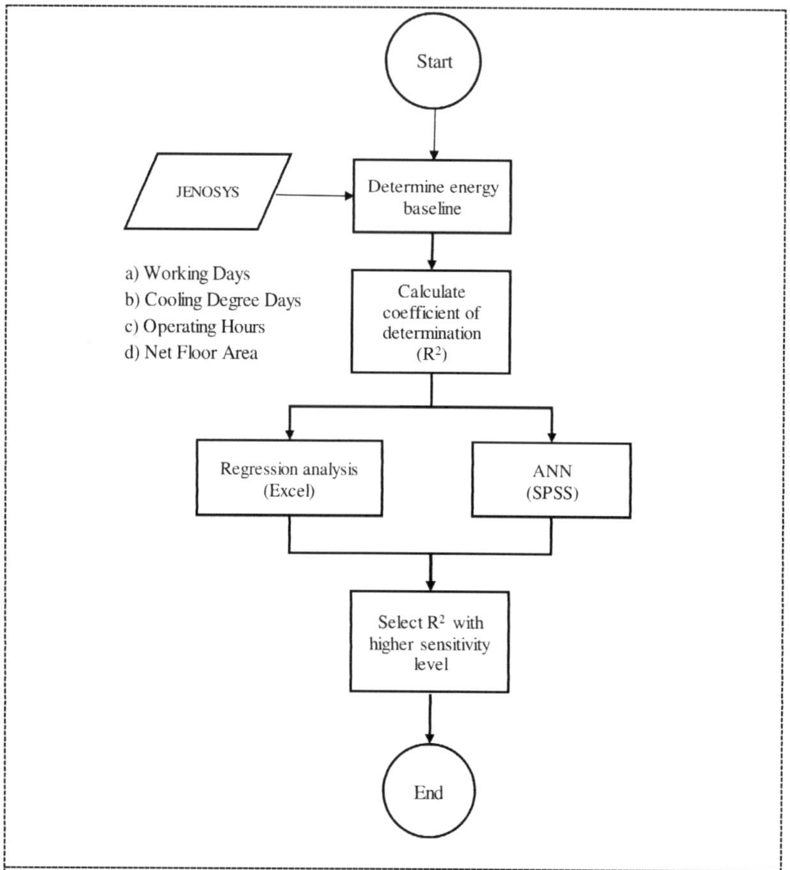

Figure 3. Baseline determination process flow for energy benchmarking.

To compare the accuracy of the models in calculating the baseline, the RA and ANN models were fed identical input parameter with static factor such net floor area and load operational hour, from which the dependent output were calculated, shown in figure 4.

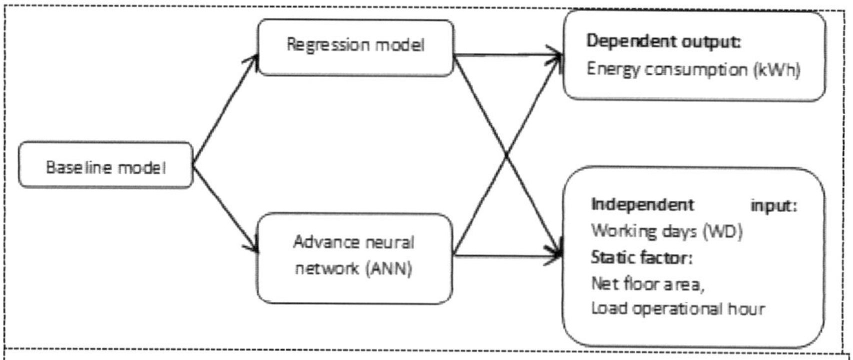

Figure 4. Variables and parameters used for ANN and regression model using the 2019 dataset of a 17-storey office building in Kuala Lumpur (Table 2).

A reliable assessment of ANN prediction accuracy is critical in selecting an appropriate architecture. The feed-forward multi-layer perceptron (MLP) architecture is used in this paper by using SPSS software, the architecture of which is shown in figure 5. The input neurons are just the same number as the ANN inputs, and because the ANN only has one output, only one output neuron is chosen. For the ANN using the parameters taken from energy audit exercise, the selected variables are working days (WD), cooling degree days (CDD), working hours, and net floor area, all of which are used as input layers. Layers 1 and 2 will always regress to find the best fit. The output layer for the ANN is the energy baseline of a building. This ANN model was used to compare with the performance of RA simulated on Microsoft Excel.

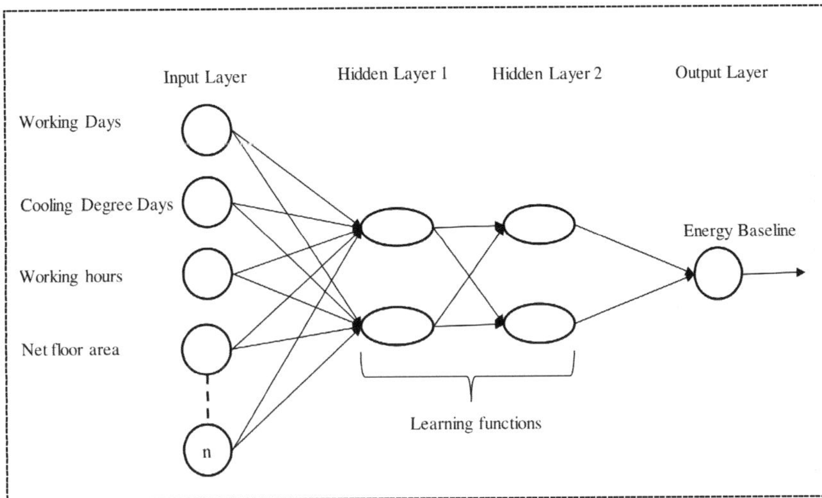

Figure 5. ANN architecture to determine the energy baseline.

The details of multi-layers in the ANN method are shown in table 1. The ANN consists of three layers which are the prediction is based on Levenberg-Marquardt algorithm (TRAINLM) as training function, Gradient descent with momentum weight and bias learning function 2 (LEARNGDM) as learning function, and Mean Square Error (MSE) and Mean Absolute Percentage Error (MAPE) as performance function.

Table 1. Multiple Layers in ANN Method.

Layer	Equation	Description (features)
Input	Levenberg-Marquardt algorithm (TRAINLM) $x_{k+1} = x_k - [J^T J + \mu I]^{-1} J^T e$	• **J** is the Jacobian matrix that contains first derivatives of the network errors respect to the weights and biases • **e** is a vector of network errors • Fastest method for training moderate-sized feedforward neural networks (up to several hundred weights).
Hidden	Gradient descent with momentum weight and bias learning function 2 (LEARNGDM) $\partial \mathrm{W} = \mathrm{MC}\,\partial \mathrm{W_{prev}} + (1 - \mathrm{MC})\mathrm{LRgW}$	• It calculates the weight change dW for a given neuron from the neuron's input P and error E, the weight (or bias) W, learning rate LR, and momentum constant MC,

		according to gradient descent with momentum: • Previous weight change dW_{prev} is stored and read from the learning state LS
Output	Mean Square Error (MSE) and Mean Absolute Percentage Error (MAPE) $MSE = \frac{1}{n}\sum_{i=1}^{n}\lvert t_n - p_n\rvert^2$ $MAPE = \frac{1}{n}\sum_{i=1}^{n}\left\lvert\frac{t_n - p_n}{t_n}\right\rvert$	• Smaller values of MSE and MAPE indicate that the results are more accurate • t_n is the targeted output data, p_n is the predicted output data from the ANN, n is the number of samples in the data set

4. Results and discussion

The comparison exercise to select the method with higher coefficient of determination (R^2) was conducted using the energy consumption, working days and cooling degree days data extracted from the 2019 energy audit report of a 17-storey office building located in Kuala Lumpur, the full set of which is shown in table 2. The building energy intensity (BEI) of the office building was calculated by dividing the total annual energy consumption of 3,546,502 kWh with the nett floor area of 37,970 m^2, which gave the value of 93 kWh/m^2/year.

Table 2. Data set used for regression model using Microsoft Excel.

	Energy consumption (kWh)	Working days (WD)	Cooling degree days (CDD)
January	300,745	21	132
February	250,971	17	131
March	306,248	21	159
April	319,360	22	143
May	299,357	20	155
June	263,852	18	135
July	317,391	22	140
August	306,071	21	138
September	276,491	18	139
October	308,875	22	96
November	300,757	21	105
December	296,384	21	107
Total	3,546,502	244	1580

The comparison of R^2 between RA and ANN at a single iteration is shown in table 3, with the best fit for ANN is shown in figure 6.

Table 3. Comparison of sensitivity between RA and ANN.

	Layer	Number of Iteration	Coefficient of Determination (R^2)
Regression Model	Single	1	0.93
ANN	Single	1	0.95

It has been shown that the ANN model has a higher sensitivity level than the RA given the same independent factor, dependent factor, static factor, and dataset size. This suggests that by using ANN, standard baseline models for government buildings can be developed more accurately based on given data.

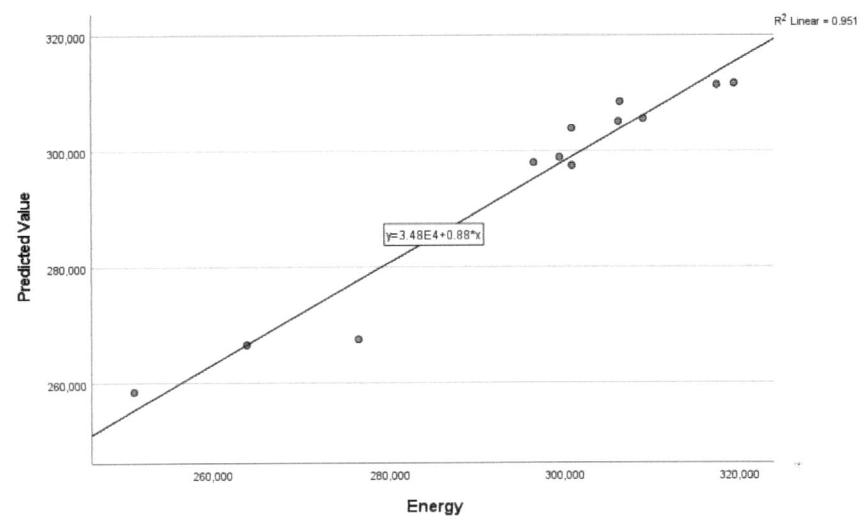

Figure 6. Coefficient of determination for ANN.

The data set from table 2 and the equation in figure 6 were used to calculate the mean absolute percent error (MAPE), the formula of which is shown in table 1. The MAPE was calculated to 0.8% as shown in table 4.

Table 4. Calculation of MAPE for ANN.

	Actual value (kWh)	Predicted value (kWh)	Absolute percent error
January	300,745	299,456	0.43
February	250,971	255,654	1.87
March	306,248	304,298	0.64
April	319,360	315,837	1.10
May	299,357	298,234	0.38
June	263,852	266,990	1.19
July	317,391	314,104	1.04

August	306,071	304,142	0.63
September	276,491	278,112	0.59
October	308,875	306,610	0.73
November	300,757	299,466	0.43
December	296,384	295,618	0.26
	Mean absolute percent error (MAPE)		0.8

The next step of the study is to add more parameters such as ambient temperature to the layers in ANN to ensure the energy can be monitored more accurately, and to verify the performance of e-EBS by simulating the energy performance and the green rating for other categories of buildings, such as mosques and university buildings. In terms of programming, the ANN model would be further developed by adding input layers, such load operating hours, or net floor area, to improve the energy baseline, the process of which is shown in figure 7.

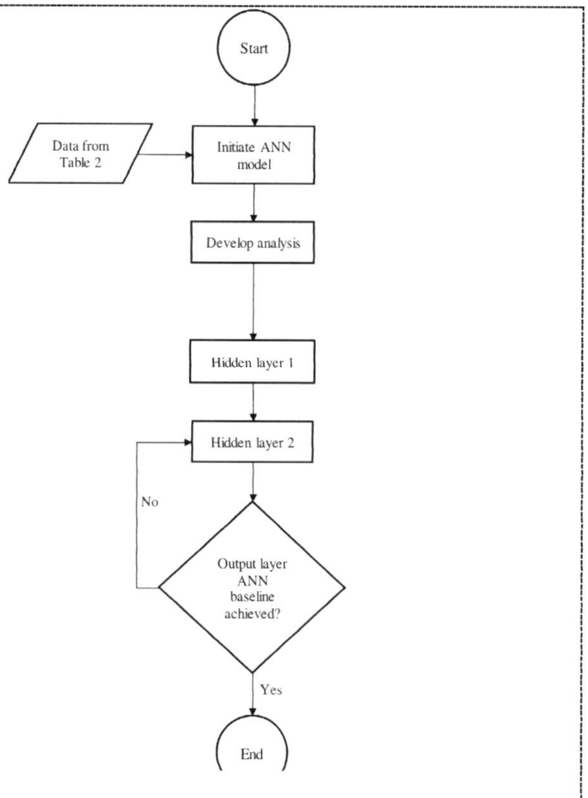

Figure 7. Flowchart for ANN analysis for energy baseline.

At the time of writing, efforts are being made to integrate the e-EBS dashboard with JENOSYS and data obtained from energy audit report, the layout of which is shown in figure 8. The dashboard displays a range of menus that contains a list of organisations, buildings as well as types of building services.

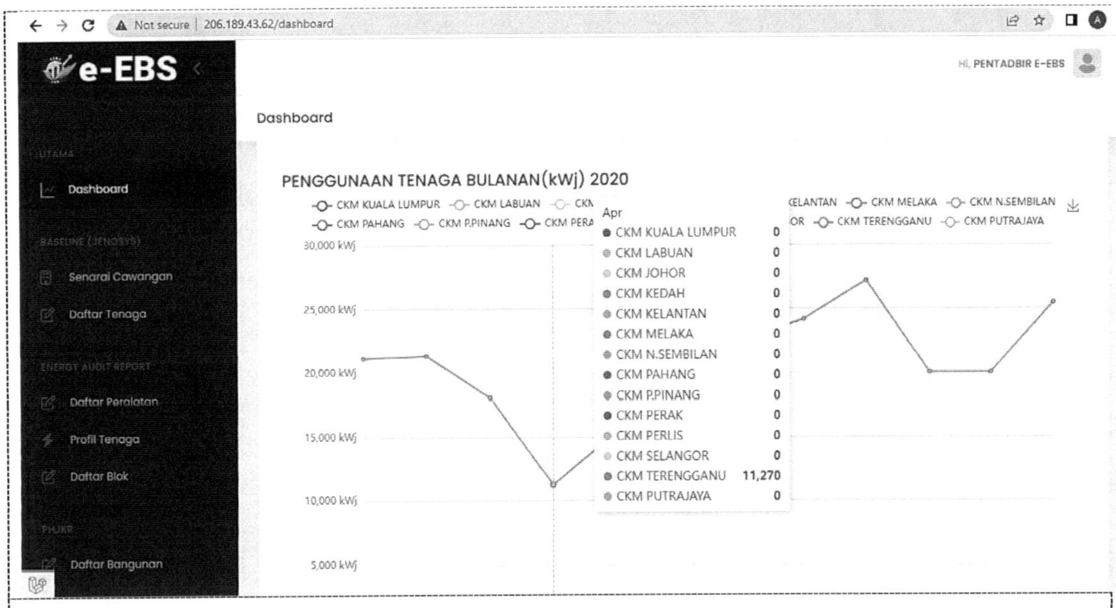

Figure 8. Layout of the e-EBS dashboard displaying monthly energy consumption for buildings.

5. Conclusion

It is proposed that ANN is used to determine adjusted baseline energy for quantifying energy savings from an ESM implemented in PWD HQ building blocks in Year 2013- 2019. The input data to the ANN includes the number of working days (WD) and cooling degree days (CDD) cumulatively for 1 year (from year 2013- 2019) period before implementation of the retrofitting program. On the other hand, output data is baseline energy use (i.e., before ESM measures). Once the optimum training parameters have been obtained, adjusted baseline energy is determined by feeding the number of WD and CDD in the chosen ESM period to the network. The performances of the ANN model are then compared with multi-regression technique in terms of coefficient of determination (R^2).

The potential applications of e-EBS are as follows, namely a) the system can be used to recommend further energy saving measure for the building owners to increase the overall green building rating, b) it can be deployed as a decision-making tool for energy policy makers and solution provider to building owners to plan a more structured energy management programme, and c) it can be used as a comprehensive tool for energy consumption by benchmarking for pre-assessment purposes.

6. References

[1] Tsemekidi Tzeiranaki S, Bertoldi P, Diluiso F, Castellazzi L, Economidou M, Labanca N., Ribeiro Serrenho T and Zangheri P 2019 *Energies* **12**
[2] Ding Y, Wang Q, Wang Z, Han S and Zhu N 2019 *Energy and Buildings* **202**
[3] Abd Rahman N M, Lim C H and Fazlizan A 2021 *Journal of Building Engineering* **43** 103088
[4] Ab Wahab N, Mat Yasin Z, A Salim N and A Aziz N F 2020 *Indonesian Journal of Electrical Engineering and Computer Science* **17**
[5] Kamaluddin K A, Imran M S and Yang S S 2016 *Journal of Building Performance* **7**
[6] Nattenee T, Wongkot W and Det D 2017 *Energy Procedia,* **141** 255-9
[7] Ridwana I, Nassif N and Choi W 2022 *Buildings* **10** 198Mellit A and Pavan A M 2010 *Energy Conversion and Management* **51** 2431-41
[8] Xin Y 1999 *Proceedings of The IEEE* vol 87 p 1423

[9] Carotenuto P L, Della Cioppa A, Marcelli A and Spagnuolo G 2015 *Neurocomputing* **170** 393-405

[10] Mat Wajid N, Zainal Abidin A M, Fadzil F and Mukhtar M Y A 2018 *International Journal of Energy and Power Engineering* **12**

Acknowledgments

The authors wish to thank the Public Works Department Malaysia for the financial support of the research and the University of Malaya for providing the academic expertise and facility.

Mobile Learning to promote Renewable Energy Education at the Secondary Education level in developing countries.

Fahimul Hoque[1], Ruhizan M Yasin[2*] and Kamaruzzaman Sopian[1]

[1]Solar Energy Research Institute (SERI), Universiti Kebangsaan Malaysia, Bangi 43600, Malaysia;
[2]Faculty of Education, Universiti Kebangsaan Malaysia, Bangi 43600, Malaysia

* Corresponding author: ruhizan@ukm.edu.my

Abstract. Education is a fundamental component in advancing society towards a cleaner and sustainable energy consumption route. To achieve and promote the Sustainable development Goals (SDG's) education will play a pivotal role in creating awareness and promotion of Renewable energy concepts amongst the general public. Malaysia and Bangladesh are two ambitious countries that has implemented various policies to harness energy from renewable sources but lacked to achieve their targets. This is mainly due to lack of RE education and interest and awareness amongst the public. Recent studies have shown a significant lack of RE curriculum specially in the secondary level of education in most of the developing countries. This lack of curriculum can directly impact the societal process to shift to a sustainable green energy future due to the lack of interest, awareness in students for RE career and importance. This gap of introducing RE education into the secondary education curriculum can be fulfilled by using Mobile Learning (M-Learning) methods and technologies as means to educate secondary schools students about RE. Both the countries have a population well equipped with smart devices and internet access thus equipping people with greater mobility, communication, and access to resources all over the world via the Internet, which can be used potentially to incorporate within the education sector as a mean for teaching and learning. Thus, this paper aims to highlight the educational strategies, theories and components that are associated with mobile learning and propose a mobile learning system that can be used to incorporate RE education in secondary education level in countries like Malaysia and Bangladesh.

Keywords: Renewable Energy Education, Mobile-learning, Sustainable Development Goals, Secondary Education, Developing countries.

1. Introduction

In the last decade, the world has shifted to a more technologically advanced society with a rapid increase of mobile technologies available and adopted by the public in the form of smart phones, tablets or PDAs (Personal Digital Assistant). This shift in technological advancement is no different in countries like Malaysia and Bangladesh where the former aims to be a fully-fledged developed country by 2024-28 and the latter aiming to be promoted to a developing country by 2026[1], [2]. As members of the United Nations both the countries in focus are part of the global Sustainable Developments Goals (SDGs) plan

set up in 2015. The SDGs consists of 17 interlinked global goals for a sustainable future for all and amongst them investment and transition to Renewable Energy (RE) sources for energy harness is a fundamental part of the SDGs. Like other countries Malaysia introduced various government policies to achieve the RE SDGs, like the 9th Malaysian Plan, Renewable Energy Act 2011 by which Malaysia aims to harness 30% of its energy needs from RE sources by 2030 and 50% by 2050 [3]. But in reality, Malaysia only managed to achieve only 2% of energy from RE sources. On the other hand, Bangladesh issued the Renewable Energy Policy in 2008 aiming to achieve 10% of its energy needs from RE sources by 2020, inviting public and private investment in RE ventures [4]. But this plan was well short of its initial target only achieving 3.5% by the end of 2020. The government issued a new plan of achieving 40% of energy from RE sources by 2041 but doubts have already risen about this if drastic measures are not taken [4], [5].
In both the countries the common factor in not achieving their respected goals was the lack of awareness, availability of knowledge on RE concepts amongst the general public [4], [6]. Education plays a pivotal role in creating and promoting awareness about various concepts that in turn helps drive the society to achieve their goals. Similarly, RE education is vital for a society to shift to clean and sustainable green energy from fossil fuels if they are implemented early in the secondary school level. Students can comprehend the RE technologies and their concept if implemented in the early stages and gain motivation and interest to pursue higher education and career in this sector [7], [8]. Thus, with the latest technological advancement available inboth the countries and the lack of RE awareness and education, this paper aims to highlight the educational strategies, theories that are associated with mobile learning (M-learning) and how they canbe used to incorporate RE education in secondary education level in both the countries.

1.1. Mobile learning in Malaysia
The rapid advancement in mobile technologies seen throughout the world in the last decade is impacting the educational sector in a positive way, with M-learning exhibiting immense potential of learning and effective teaching, thanks to the advantages mobile technologies have with them like portability, participatory learning, ease of communication and availability [9]. According to Statista 2022, 88.79% of the total population of Malaysia use smartphones [10] and the Department of Statistics Malaysia published a report showcasing 96.85 of the population have internet access [11]. Amongst the population that uses smartphones 13.5% belongs to age group 15-19 and 24.5% comprises of the age group 20-24 (refer to Figure 1), thus highlighting a high number of secondary school students and tertiary level students owns smartphones that can be used for mobile learning in their respected institutions [12].

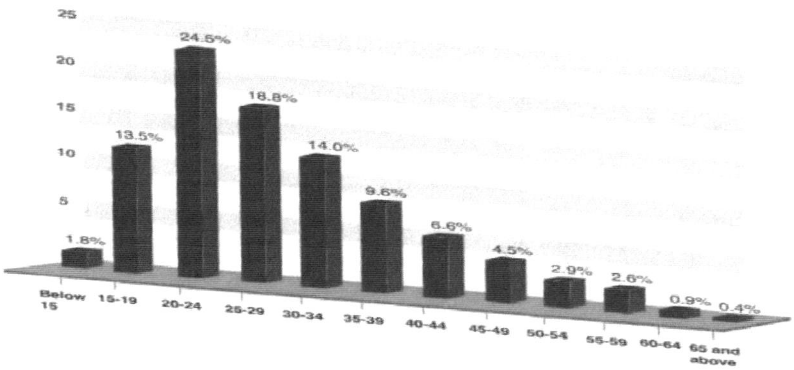

Figure 1: Percentage of Smartphone Users by Age Category [12]

Government intervention to implement M-learning in the educational spectrum of Malaysia is critical to bring about the success it is capable of bringing. The government with the help of dedicated policies that can enable M-learning in various institutes across different states can help students access good quality education and increase their technological literacy [3],[7]. The Malaysian government has already implemented various policies like "National Higher Education Strategic Plan" beyond 2020, which emphasizes on the usage of ICT and mobile technologies as means to offer effective learning and teaching in higher education institutions in Malaysia, and also the "Bring your own device" (BYOD) policy where teachersand students bring their own portable and use the institutions network system to convey teaching and learning [13]. These policies have seen a positive impact in providing quality education to all especially in the rural areas with help of remote learning. But the setback is the lack of effortgiven to implement these policies in the secondary level of education [13] which if implemented can help create awareness, interest and motivation for these students to pursue higher education and career [14], [15] in necessary areas like the RE sector which is in need of better-quality workforce and innovation.

1.2. Mobile learning in Bangladesh

With the advancement in technological innovation and the current digital age the educational sector has seen massive changes in their strategies to incorporate the 21st Century skills characteristics. In the current educational trend across the world students are expected to master the 21st century skills like knowledge construction, skilled communications, collaborations with peers, technological literacy to name a few, as they are seen as necessity characteristic requirement in modern workforce [16]. Thus, alongside Malaysia and other countries Bangladesh also implemented policies to promote M-learning and ensuring the students are equipped with these skills and characteristics required in the modern workforce industry. The government of Bangladesh implemented the National Information and Communication Science policy in 2009, which emphasizes the use of digital and mobile technologies for ICT based education across primary to tertiary level of education. This policy is part of the greater national goal of "Digital Bangladesh: dreams and reality". Bangladesh being one of the fastest smartphone mobile penetration countries in theworld, with 47% of its total mobile phone users owns a smartphone [17]. Around one million of the smartphone users falls in the age group of 18-25, and a possible significant number can be found in the15-18 age group, thus encompassing a huge number of secondary and tertiary level of students that canimplement M-learning for effective learning and teaching methods [18]. Under the aforementioned policies the government has taken great steps in providing M-learning infrastructure like free wi-fi in all public universities, so that teachers and students can use the network to implement M- learning [18], [19]. But similar to Malaysia, there is a lack of effort and infrastructure present in the secondary level of education. This is because the policies currently implemented cater to the higher education spectrum and less policies are implemented in the secondary education [3],[4],[6]. Thus, this paper aims to highlight the importance of using M-learning to provide quality education for all including students in the remote areas of the country and in turn use M-learning to promote RE education amongst the secondary school students.

2. Mobile learning characteristics and educational theories related to M-learning

Mobile learning has been defined by Crompton, Berge and Muilenberg in their book *"A Historical Overview Of M-Learning"* as *"the process of learning across multiple contexts, through social and content interactions, usingpersonal electronic devices"*. The term "context" in the definition means that m-learning is self-directed and spontaneous meaning learning can be done by an individual by his or herself or by directions of others in a spontaneous learning environment [20]. Thus, based on the definition m-learning can be divided into elements that forms the basis of this learning method. The elements include the learner which is more student centered based on the 21st century learning and teaching model, where the learner need to have access to the information at any time and are able to self-evaluate their learning as individuals or in collaboration with peers. In m-learning the learner is at the center of the process and all other elements revolve around them. The other elements are teacher who are apart from being only providers of information needs to collaborate with students and supervise them as a moderator or consultant using modern mobile technologies to acquire knowledge [21]. Content is another element in M-learning which needs to be designed and developed in various multimedia formats ranging from animation graphics, interactive videos, quizzes, or interactive games to allow learners to quickly absorb the information using the technologies [21], [22]. Environment is another element of M-learning that must be designed in a way to promote positive learning experience, either in the form fully online, face-to-face, or mixed mode. Depending on the mode, the learners should always have access to the information required and have the ability to communicate and collaborate with fellow students and teachers to achieve their learning goals [21], [22]. And lastly assessment is the last element of M-learning that enables teachers to monitor and analyze the learners' performances using online exams, online discussions and creating database logs to evaluate the students' progress. M-learning emphasizes on the ability of students to self-evaluate their progress by conducting online tasks, quizzes and discussions and getting feedback instantly via the application, website or devices they are using [23].

2.1. *Characteristics of M-learning*

To implement M-learning model and develop a learning system in the form of website or application, the characteristic of the model needs to be followed and kept in mind while designing the system. The core characteristics allows the learners to experience an efficient learning environment with ease of access to required knowledge at any time and place [24]. The core characteristics are Ubiquitous and Portability which allows learners to access information from anywhere without a fixed geographical location and also allows learners to use portable wireless devices to perform learning tasks and activities anytime from anywhere. Blended is another characteristic that allows instructors to use various teaching methods and strategies to provide the necessary learning outcomes for the learners. With the technology present the model allows for blended learning which is beneficial to conduct a mixture of face-to-face and online learning process [21], [25], [26]. Interactiveand Collaborative are two characteristics that allows learners to interact with the system to gain knowledge and also enables them to communicate with fellow peers and teachers in a social construct to complete tasks and get feedback instantly creating a collaborative learning environment [23], [27]. Instantaneous and Private information are the last two fundamental characteristics as they provide the learners with safe and secure learning environment where they can get private feedback on their progress and can access all the contents in the system at any time or place [21],[27]. These characteristics are fundamental in providing a successful learning environment for the learners and this paper uses these characteristics to come up with the design of a M-learning system forRenewable Energy education.

2.2. *Educational learning theories related to M-learning:*

The way each learner gains knowledge varies differently from person to person, thus many researchers across the world conducted research to understand the different ways learners gain knowledge. Based on these researches they came up with learning theories that can be implemented alongside educational teaching methods and strategies to provide efficient learning opportunities to learners. Similarly, M-learning methods also corresponds with some of these learning theories to makeit a more efficient learning method [28]. The learning theories that form the core basis of M-learning are Constructivism which is credited to John Dewey as being the founding philosopher. And Social Constructivism founded by Lev Vygotsky [29].Constructivism theory states that leaners acquire knowledge themselves based on their past experiences.The key features of this theory are that the learning process should be active where the learners aregiven meaningful activities requiring high level of processing. The knowledge that the learners gain should be constructed by the learners themselves and the teacher should be more like a facilitator or moderator, giving interactive instructions to encourage the learners. The learners must be motivated to be involved in collaboration and cooperative learning making the learners work with fellow peers and instructors to gain real-life experiences and perform metacognitive skills to execute the given tasks. The theories firmly states that learners should be at the center of the learning process, and they shouldindulge in meaningful interaction to promote high-level of social learning process [30]. Also, social constructivism's core belief is that knowledge is developed by learners through social interactions withpeers and instructors. Vygotsky proposed the concept of "Scaffolding" in his theory which is a process where students are supported by teachers, peers, or adults to gain knowledge about new concepts through meaningful social interactions. Both the theories emphasizes mostly on collaborative learning where learners gain knowledge working in groups [28], [30], [31]. These are justtwo of the many learning theories that can be applied to M-learning and in this paper these two theoriesare the basis on which the M-learning system for Renewable Energy education was designed.

3. Proposed design of the M-learning system for RE education

With the current availability and usage of mobile devices, like laptops and smartphones, web applications have gained immense popularity amongst the users. Web applications are different from websites, where websites can only be used by using a browser on the laptop and mobile phones, web applications can be used by browsers in the laptop and also as a mobile application in smartphones making it more dynamic and easier to for developers to make constant updates and allow users smooth and efficient user experiences [32]. The proposed design of the M-learning system for RE education is a web application with this network architecture designed in such a way that allows users portability and also security from external attacks. The personnel responsible for maintaining and providing updatesto the users will be stationed at the institutions' offices protected by a firewall from malicious attacks and the data components will be stored in a registered online server so that the system is accessible by the users via their laptops or smart devices at any time (refer to Figure 2).

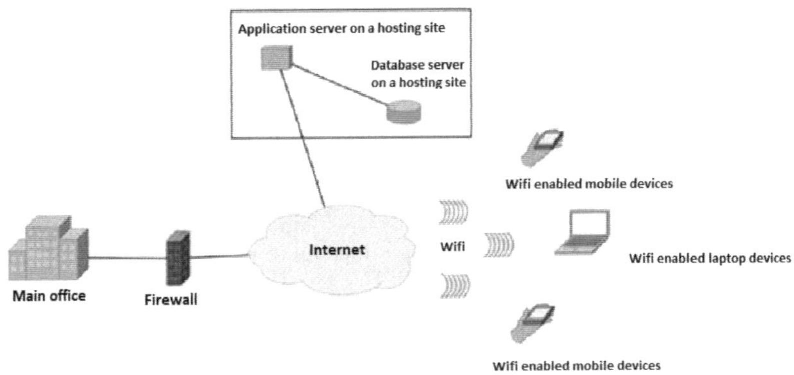

Figure 2: Network Architecture Diagram for the proposed M-learning system.

The system's functionalities are divided into three categories based on the three different types of users that will be using the system. The first is "Admin" who is responsible for approving registration of teachers by checking the credentials provided during registration. He or she is also able to see all the contents and discussion present in the system and make necessary changes to remove any harmful contents or discussion languages. The admin will generally be a member of the institution's staff thus he or she also has the ability to upload comments or post notices on behalf of the institution if required. Second type of user are the teachers, who are able to register and login to the system once their registration is approved. They can upload contents which can be in the form of multimediaslides, videos, or articles. The contents preferred to be in the form of a combination of texts, images, graphics alongside animation or interactive videos are widely accepted to be a better form of deliveringinformation to the modern generation who are more used to this type of media in their daily lives [33], [34]. The teachers can create quizzes for the student to take part in and upload assignments or activitiesto assess the student's progress. They can also engage in the discussion forum alongside students in discussing issues and providing guidance to the students. Finally, the third type of users are the studentswho can access all the uploaded contents, engage in the various threads in the discussion forum with fellow peers and instructors and participate in the online quiz and activities to enhance their knowledge and also get real time feedback on their assessment from the quiz scores (refer to Figure 3).

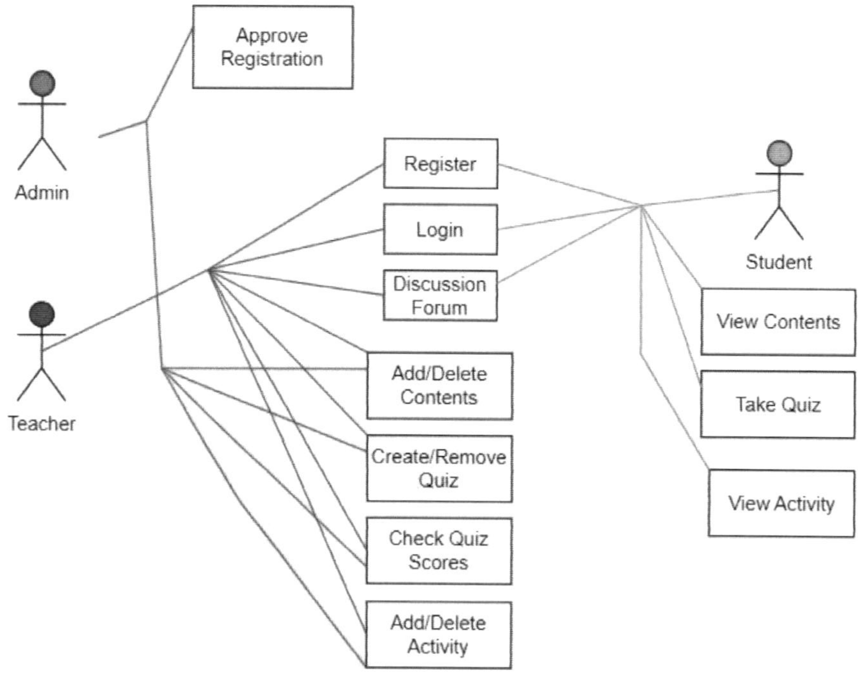

Figure 3: Functionalities of the proposed system

4. Discussion

The proposed M-learning system was designed keeping in mind all the characteristics of M-learning and all the educational strategies and theories that are closely associated with M-learning, to give the learners a positive environment to gain knowledge. Since the core characteristics of M-learning includes Ubiquitous and Portability, the network architecture of the system is designed in such a way that the system is always kept online and accessible to the users via any portable device at any time. The system being proposed as a web application rather than just a website allows more flexibility to the users to use it anytime from laptops or PCs via browsers or as mobile applications in their Android, iOS or Google operating system devices (refer to Figure 2). The functionality named "Discussion forum" (refer to Figure 3) allows students and teachers tohave communication with each other via posts and comments method at any time, allowing collaborative learning amongst fellow students. This functionality encompasses two of the core characteristics of M-learning "Interactive" and "Collaborative" and keeps in mind the educational theory of Constructivismand Social constructivism, where students form social interactions with each other and also with instructors to gain knowledge. Instantaneous and Private are two other characteristics of M-learning thatheavily influenced the proposed system's design. With various studies showing great success in implementing "Anytime, anywhere learning" which allows students in remote areas to high quality education [35], [36], this system also allows its users to access the contents at any time and any place instantly and provides security of their data. Blended is the last characteristic of M-learning that the proposed system adheres to encouraging and allowing teachers to conduct both face-to-face and onlinemethods of teaching. The teachers are allowed to upload digital contents in the system and teach them in class and later allow students to do additional reading on the contents at home and take part in quizzes online. The students are allowed to take quizzes at any time and the result of the quiz will be given instantly providing instantaneous assessments of their progress, which gives motivation and encouragement to students on achieving their educational goals. Also, teachers can upload activities to be done in group or individually which students can do at their own time and upload to the system for assessment. Formative assessment (providing feedbacks to learners on current performances) [37] can be given by instructors via the discussion forum to students in the form of private or group discussions with students. And summative assessment (providing feedback on the learning process at the end of a task or activity) [37], [38] can be provided by instructors to the learners in the form of activity marks or also from total quiz scores. The functionalities of the admin to approve registration and the authority to remove any contents and posts and comments is given to maintain security and integrity of the overall system.

Another important aspect of the system that needs to be addressed is the type of content like digital media (multimedia slides, videos, animations) that needs tobe uploaded for the students to use and learn. With advance technological environment the current generation are well equipped to absorb information from digital media like animations, videos, or image-based slides. Studies have shown digital media impacts positively on the student's information retentionand performance if equipped with metacognitive and motivational prompts from the instructors [39], [40]. Thus, the proposed system allows teachers to upload slides on the topics of RE and its concepts and technologies, in the form of text as well as images or graphics, videos either as interactive or animations and also extra readings in the form of articles, multimedia forms of books or documents. Thecontent of the system should be multimedia and digital based for the overall system to succeed in achieving its goal of RE education.

4.1. Suggestions for Teachers and Policy makers

Teachers play a pivotal role in implementing M-learning. Apart from being providers of information in M-learning setting teachers needs to play active part in providing instructions and motivational prompts regularly to the students. This encourages the students to complete their tasks and necessary activities. For the proposed system teachers are needed to have regular interaction with the students using the discussion forum. The teachers also need to encourage the students to use the functionality to regularly communicate and collaborate with fellow students and teachers. Teachers also need to provide regular prompts and tasks/activities using the system so that the students are part of an active learning environment. Teachers also need to provide assessments individually to the studentsto create more motivation and guidance to learn the RE concepts.

In previous research, M-learning has shown positive impact on students' knowledge, motivation and awareness on particular subjects. According to a study conducted in secondary schools in Greece, using mobile learning method of instruction students have shown to display increase in interest, knowledge in science subject and physics experiments [41]. In various countries researchers have conducted experiments to investigate the impact of using M-learning methods for particular subjects like science, geography, physics and have gotten positive results in increasing students' knowledge, awareness about the topics and also their motivation to pursue further education on the selected subjects [41], [42]. Thus, policy makers in both the countries, Malaysiaand Bangladesh, should implement drastic measures to make M-learning for RE education accessible for all the students in secondary schools, so that the future prospective students can be encouraged to pursue higher education and career in the RE fields [43]. Policy makers should implement teacher's training and increase teachers' technological literacy [13] to make it easier and effective for them to usethe M-learning technologies and establish secure network infrastructure in schools to allow students to use the network to conduct their tasks using the system [44], [45].

5. Conclusion

Malaysia and Bangladesh, both have high ambitions to achieve a sustainable green energy future for their citizens but are lacking awareness, education, and motivation amongst their public to achieve their desired RE goals. To overcome these hurdles, educating and creating awareness and interest amongst the public has become fundamental for both the countries, specially to produce future generations of people interested in pursuing careers in RE sectors and generate more technologically advanced workforce, to achieve their long term RE goals. Thus, with both the countries having vast number of students who are technologically well adapted, implementing M-learning methods to educate these students on RE concepts and practices will come a long way in achieving each country's respected RE goals, with the help of proper intervention from the policy makers in conducting teacher's training on technological literacy and also developing network infrastructures for the students to securely use them for their educational goals.

Funding and Acknowledgement:
This research was funded by the "Ministry of Higher Education, Malaysia", (Grant number LRGS/1/2019/UKM-UKM/6/4). We would like to thank the Ministry of Higher Education Malaysia, Faculty of Education, UKM and SERI UKM for providing support for the study.

References

[1] R U M Rejaul Karim Byron, Feb. 28, 2021 "Becoming A Developing Nation: Bangladesh reaches A Milestone," *The Daily Star*, Dhaka.
[2] KWSP Mar. 21, 2021 "Malaysia to achieve a transition to high-income economy in 2024-2028 – World Bank,".
[3] H Azil Illias, N Syuhada Ishak, N Aini Mohd Nor Alam, S Menengah Kebangsaan Assunta, J Changgai, and P Jaya 2020 "Awareness Towards Renewable Energy Among Secondary School Students in Malaysia," *INT. Journ. of Ren. Ener. Research.*
[4] M E Karim, R Karim, M T Islam, F Muhammad Sukki, N A Bani, and M N Muhtazaruddin Oct. 2019 "Renewable energy for sustainable growth and development: An evaluation of law and policy of Bangladesh," *Sustainability (Switzerland)*, **11**, doi: 10.3390/su11205774.
[5] Eyamin Sajid Sep. 05, 2022 "Regional energy trade: This is how Bangladesh can achieve renewables target," *The Business Standard.*
[6] N F Khairudin, R Bidin, A H Shamsuddin, A Akhiar, F Ideris, and A A Rahman Jun. 2020 "Renewable energy development in Malaysia: Communication barriers towards achieving the national renewable energy target," in *IOP Conference Series: Earth and Environmental Science*, **476** doi: 10.1088/1755-1315/476/1/012080.
[7] I Ocetkiewicz, B Tomaszewska and A Mróz 2017 "Renewable energy in education for sustainable development. The Polish experience," Renewable and Sustainable Energy Reviews, **vol. 80**. Elsevier Ltd, pp. 92–97. doi: 10.1016/j.rser.2017.05.144.
[8] T C Kandpal and L Broman 2014 "Renewable energy education: A global status review," *Renewable and Sustainable Energy Reviews*, **34**. Elsevier Ltd, pp. 300–24. doi: 10.1016/j.rser.2014.02.039.
[9] N Aina Syakirah Zainal Abidin and S Wei Tho 2018 "The development of an innovative resonance experiment using smartphones with free mobile software applications for tertiary education"
[10] Statista Research Department Oct. 2022 "Number of smartphone users in Malaysia from 2010 to 2020 and a forecast up to 2025," Hamburg.
[11] Department of Statistics Malaysia Apr. 2022 "ICT USE AND ACCESS BY INDIVIDUALS AND HOUSEHOLDS SURVEY REPORT 2021" Putrajaya.
[12] S M Shuhidan, S R Hamidi and I S Saleh Aug. 2017, "Perceived Risk towards Mobile Banking: A case study of Malaysia Young Adulthood," in *IOP Conference Series: Materials Science and Engineering*, **226**. doi: 10.1088/1757-899X/226/1/012115.
[13] Nor Fadzleen Sa'don, Halina Mohamad Dahlan and Abdelrahman Ibrahim 2014 "Usage of Mobile Learning in Malaysian Secondary Education: Stakeholders' View," *Journal of Information Systems Research and Innovation*, **6**, pp. 42–50.
[14] H. Korkmaz 2015 "Factors Influencing Students' Career Chooses in Science and Technology: Implications for High School Science Curricula," *Procedia Soc Behav Sci*, **vol. 197**, pp. 966–972, doi: 10.1016/j.sbspro.2015.07.284.
[15] A D Madden, S Webber, N Ford and M Crowder 2018 "The relationship between students' subject preferences and their information behaviour," *Journal of Documentation*, **vol. 74, no.**

4, pp. 692–721, doi: 10.1108/JD-07-2017-0097

[16] S M Thang, L Y Sim, N Mahmud, L K Lin, N A Zabidi and K. Ismail March 2014 "Enhancing 21st Century Learning Skills Via Digital Storytelling: Voices of Malaysian Teachers and Undergraduates," *Procedia Soc Behav Sci*, **vol. 118**, pp. 489–494, doi: 10.1016/j.sbspro.2014.02.067.

[17] Halima Binte Islam 2022 "Smartphone users will grow to 63% by 2025: Report," *The Business Standard*, Dhaka.

[18] B Biswas, S K Roy and F Roy 2020 "Students Perception of Mobile Learning during COVID-19 in Bangladesh: University Student Perspective," *Aquademia*, **4-2**, p. 200-23, doi: 10.29333/aquademia/8443.

[19] M Syed Ahmed and S M A Kabir Apr. 2018 "The Acceptance of Smartphone as a Mobile Learning Tool: Students of Business Studies in Bangladesh," *Malaysian Online Journal of Educational Technology*, **vol. 6, no. 2**, pp. 38–47, doi: 10.17220/mojet.2018.02.003.

[20] H Crompton 2013 "A historical overview of mobile learning: Toward learner-centered education," in *A Historical Overview of M-learning*, Z. Muilenberg, L. Berge, and L. Y., Eds. New York: Routledge, pp. 3–14. doi: 10.4324/9780203118764.ch1.

[21] F Ozdamli and N Cavus 2011 "Basic elements and characteristics of mobile learning," in *Procedia - Social and Behavioral Sciences*, **vol. 28**, pp. 937–942. doi: 10.1016/j.sbspro.2011.11.173.

[22] L Siragusa, K C Dixon, and R Dixon 2007 "Designing quality e-learning environments in higher education,".

[23] M Sharples, J Taylor, G Vavoula, and V A Giasemi "A Theory of Learning for the Mobile Age." [Online]. Available: https://telearn.archives-ouvertes.fr/hal-00190276

[24] P Seppälä and H Alamäki 2003 "Mobile learning in teacher training,"

[25] N Cavus and D Ibrahim 2009 "M-Learning: An experiment in using SMS to support learning new English language words," *British Journal of Educational Technology*, **40** pp. 78–91, doi: 10.1111/j.1467-8535.2007.00801.x.

[26] M Ehmet and A Ocak 2010 "Blend or not to blend: a study investigating faculty members' perceptions of blended teaching,".

[27] N Cavus and H Uzunboylu 2009 "Improving critical thinking skills in mobile learning," *Procedia Soc Behav Sci*, **vol. 1, no. 1**, pp. 434–38, doi: 10.1016/j.sbspro.2009.01.078.

[28] M Sharples, J Taylor, and G N Vavoula 2005 "Towards a theory of mobile learning," [Online]. Available: www.infed.co.uk

[29] R Vanderstraeten and G Biesta 1998 "Constructivism, Educational Research, and John Dewey", *The Paideia Archive: Twentieth World Cong. Of Phil.* pp.34-39 http://dx.doi.org/10.5840/wcp20-paideia1998235

[30] M Ally (2004) "Foundations of Educational Theory for Online Learning" *Theory and Practice of Online learning* ed T Anderson and F Elloumi (Athabasca: Athabasca University) p. 3-31

[31] N O Keskin and D Metcalf "THE CURRENT PERSPECTIVES, THEORIES AND PRACTICES OF MOBILE LEARNING," 2011.

[32] F Shahzad 2017 "Modern and Responsive Mobile-enabled Web Applications," in *Procedia Computer Science*, **110**, pp. 410–415. doi: 10.1016/j.procs.2017.06.105.

[33] T K Neo and Mai 2004 "Integrating multimedia into the Malaysian classroom: Engaging students in interactive learning,".

[34] I Rajendra and I Sudana 2018 "The Influence of Interactive Multimedia Technology to Enhance Achievement Students on Practice Skills in Mechanical Technology," in *Journ. of Phy.: Conf. Series*, **953**. doi: 10.1088/1742-6596/953/1/012104.

[35] Uvarov Alexander, Varlamova Julia, and UNESCO 2019 "Anytime, anywhere learning for improved education results in Russia_ Case study by the UNESCO-Fazheng project on best practices in mobile learning - UNESCO Digital Library," Paris, 2019. Accessed: Oct. 22, 2022. [Online]. Available: https://unesdoc.unesco.org/ark:/48223/pf0000367745

[36] S Alamri, Dr M Meccawy, B. Khoja, and H. Bakhribah May 2017 "Learning Anytime, anywhere: Benefits and Challenges of M-Learning through the Cloud," *IARJSET*, **4** pp. 37–

41, doi: 10.17148/iarjset.2017.4508.

[37] S Rahman and M A A Majumder 2015 "Is it assessment of learning or assessment for learning?" *South East Asia Journal of Public Health*, **4-1**, pp. 72–74, doi: 10.3329/seajph.v4i1.21847.

[38] G Vavoula and M Sharples 2009 "Meeting the Challenges in Evaluating Mobile Learning: A 3-Level Evaluation Framework," *International Journal of Mobile and Blended Learning (IJMBL)*, **1**, pp. 54–75 doi: 10.4018/jmbl.2009040104.

[39] S S Ang, M Orozco, D Gijbels, and P van den Bossche 2018 "Learning in the Context of Work in a Digital Age: The Use of Digital Media in Informal and Formal Learning Contexts," in *Professional and Practice-based Learning*, **21**, Springer Nature, pp. 87–101. doi: 10.1007/978-3-319-63257-5_7.

[40] M Daumiller and M Dresel 2019 "Supporting Self-Regulated Learning With Digital Media Using Motivational Regulation and Metacognitive Prompts," *J Exp Educ*, **87**, pp. 161–176, doi: 10.1080/00220973.2018.1448744.

[41] K Nikolopoulou and M Kousloglou 2019 "Mobile Learning in Science: A Study in Secondary Education in Greece," *Creat Educ*, **10 no. 06**, pp. 1271–1284, doi: 10.4236/ce.2019.106096.

[42] N Parsazadeh, R Ali and M Rezaei 2018 "A framework for cooperative and interactive mobile learning to improve online information evaluation skills" *Comput Educ*, **120**, pp. 75–89 doi: 10.1016/j.compedu.2018.01.010.

[43] F. Hoque, R. M. Yasin, and K. Sopian 2022 "Revisiting Education for Sustainable Development: Methods to Inspire Secondary School Students toward Renewable Energy," *Sustainability (Switzerland)*, **14** , doi: 10.3390/su14148296.

[44] A Habib and A S M L Hoque 2010, "Towards mobile based e-learning in Bangladesh: A framework," in *Proceedings of 2010 13th International Conference on Computer and Information Technology, ICCIT 2010*, pp. 300–305. doi: 10.1109/ICCITECHN.2010.5723873.

[45] B. Biswas, S. K. Roy, and F. Roy 2020 "Students Perception of Mobile Learning during COVID-19 in Bangladesh: University Student Perspective," *Aquademia*, **4**, p. ep20023, Jul. 2020, doi: 10.29333/aquademia/8443.

AUTHOR INDEX

Institute of Physics
Dirac House, Temple Back
Bristol BS1 6BE UK

ISSN: 1757-8981
ISBN 978-1-7138-6984-9
9 781713 869849

CURRAN ASSOCIATES INC.
proceedings.com